P ROLOGUE
What's Past is

P*What's Past is* ROLOGUE

The Personal Stories of Women in Science at the Vanderbilt University School of Medicine

Edited by Eric G. Neilson, MD

Hillsboro Press

PROVIDENCE PUBLISHING CORPORATION

FRANKLIN, TENNESSEE

Eric G. Neilson, MD, is the Hugh Jackson Morgan Professor of Medicine
in the Departments of Medicine and Cell and Developmental Biology
at Vanderbilt University School of Medicine
in Nashville, Tennessee

Copy Editor: Cynthia Floyd Manley
Division of Communications,
Vanderbilt University Medical Center
Nashville, Tennessee

Printed in the United States of America

10   09   08   07   06   1   2   3   4   5

Library of Congress Control Number: 2006922036

ISBN-13: 978-1-57736-364-4
ISBN-10: 1-57736-364-7

Cover design by Eric G. Neilson, MD

TENNESSEE HERITAGE LIBRARY

HILLSBORO PRESS
an imprint of
Providence Publishing Corporation
238 Seaboard Lane • Franklin, Tennessee 37067
www.providence-publishing.com
800-321-5692

*For future young minds looking for a place
accessible to those who believe*

Contents

Professor of Pathology, Medicine, and Pediatrics

Department of Pathology

Director, Division of Renal and Electron Microscopy

Assistant Professor of Ophthalmology

Division of Oculoplastic and Reconstructive Surgery

Department of Ophthalmology and Visual Sciences

Eric G. Neilson, MD

Foreword

The title of Eric Neilson's collection of inspiring personal stories of successful women scientists at Vanderbilt University School of Medicine—*What's Past Is Prologue*—could not be more appropriate or timelier. When I attended medical school in the late 1960s, there were only a handful of women in my class at Cornell University Medical College, only a few in the residency training programs, and even fewer on the faculty. Today, equal numbers of women and men are entering our medical school classes, and women comprise approximately one third of our faculty. However, this change by itself will not produce more women realizing what the nearly thirty women scientists writing for this anthology have accomplished. We must carefully consider factors ensuring research success and develop strategies to benefit everyone—women and men—who understand how important a career in biomedical science is to American medicine.

So, what are the common themes in these stories? These women were often the first members of their families to pursue careers in medicine and research. They were supported by family and friends. Early in their training, an instrumental individual urged them to reach for the stars. Balance and support in their personal lives also appear to be key. Many have husbands who are investigators and who provide special and keen understanding and counsel. Most have children. And each has given back, through teaching, mentoring, and organization of programs to develop those women coming along behind them. Finally, they are experts at time management, and they work very, very hard at something for which they have a true passion.

As the Dean of the School of Medicine at Vanderbilt, I have become increasingly aware of these many lessons and try to incorporate such lessons into our strategies to help all faculty be successful and avoid burnout as rising leaders.[1] As in other professions, we know the positive impact of a dual career household to the success of members of our workforce. I have also come to appreciate these challenges myself as someone who worked as a professional while being a single parent of two children. There is much we can, and should, do to address the special obstacles facing our women on faculty, including providing effective mentoring and helping to address the imbalance they often face between their work lives and their home lives.

In closing, I want to congratulate Dr. Neilson not only for compiling this excellent volume but also for his efforts as a proactive mentor and a role model for our chairs at Vanderbilt and chairs throughout the country. I also congratulate each of these successful women who have contributed so much to science and to our school. It is a privilege to work with them and to learn from their experience. Several years ago, my wife, Dr. Patricia Temple Gabbe, Professor in Pediatrics, wrote a paper entitled "Reflections on Being a Spouse in Academic Medicine."[2] A quote from that paper appears most appropriate in closing:

> In the future, more physicians in academic medicine will be pregnant and then breastfeeding. They will struggle with the desires of motherhood and a successful career. Most women will undoubtedly find the multiple roles physically and emotionally taxing, especially while raising a family. However, women in academic medicine are more likely to promote more flexible career paths and job sharing for all physicians. Those of us who have raised families and partnered with our academician husbands while maintaining our own careers should pass this message on to the next generation. In the future, the supportive academic spouse may just as often be a man as a woman.

Yes, what's past *is* prologue!

Steven G. Gabbe, M.D.
Dean, Vanderbilt University School of Medicine
Professor of Obstetrics and Gynecology

Notes:

1. S. G. Gabbe, J. Melville, L. Mandel, E. Walker, "Burnout in Chairs of Obstetrics and Gynecology: Diagnosis, Treatment, and Prevention," *American Journal of Obstetrics and Gynecology* 186 (2002): 601–12.

2. P. C. Temple, "Reflections on Being a Spouse in Academic Medicine," *Obstetrics & Gynecology* 104 (2004): 389–92.

Preface

A few months ago, about twenty-five female scientists in the Department of Medicine at Vanderbilt met over lunch to discuss the future of our research enterprise and the increasing need for more women to join in the work of discovery. While women in large numbers are entering graduate school, clinical medicine, residency, and fellowship, surprisingly few are eager for careers in academic research. A litany of reasons why quickly dominated our conversation. This vexing issue has not received much thought, nor it is clear what can be done to address it. Is a real love of science undervalued by admissions committees? Is the training paradigm for women not calibrated correctly or the challenge of being an investigator too enigmatic to contemplate? Is the right mentor critical? Is the tenure clock family friendly? Many questions were raised that day that lack good answers.

We finished lunch with the sense that many women considering graduate or postgraduate training could not easily visualize laboratory life, nor identify with the academic germline that creates new information others will use. Ironically, everything in the practice of medicine we take for granted today started somewhere in basic science or translational research. So how do young colleagues connect with the scientific energy in their environment? Successful investigators usually point to the encouragement of a mentor at the moment they began to consider for themselves a career in science. Mentorship provides a mechanism for "realizing the dream," and mentors select their protégés through interpersonal resonance guided by genuine promise. The hard work of finding an important research problem and the creativity in following a big idea all matter, as does balancing personal life with the excitement of discovery. But there is more to it than that. Few of us enter any career without privately testing the choices of others.

Of the many thoughtful ideas raised by our conversations, the one that generated most enthusiasm was the notion of sharing the personal experiences of successful women scientists at Vanderbilt who had found their way. In the pages ahead is their anthology of earned wisdom. It is my hope that young women arriving at the Vanderbilt School of Medicine who want to sense the joy of discovery will be emboldened by those who have gone before . . . what's past is prologue. The women in these stories are future mentors and friends, an invaluable resource for new learners searching for their own path to scientific inquiry.

I want to thank Cynthia Manley for her help in copyediting this manuscript, Suzanne Alexander for her organization, Hillsboro Press for publishing, and of course, all the contributors for sharing the stories that make this book come alive. I was inspired by everyone's willingness to contribute something special. The reminiscences in this anthology are arranged without method. Pick through as suits; each story is unique, but all echo a common theme—you, too, can do this.

Eric G. Neilson, MD
Hugh Jackson Morgan Professor of Medicine
Professor of Cell and Developmental Biology
Chair, Department of Medicine
Vanderbilt University School of Medicine
Nashville, Tennessee
14 June 2005

P*What's Past is*ROLOGUE

Chapter 1:
NANCY J. BROWN, MD

Robert H. Williams Professor of
Medicine and Pharmacology

My laboratory conducts vascular research in humans. In particular, we study two of the body's blood pressure regulatory systems, the renin-angiotensin-aldosterone system and the kallikrein-kinin system. The renin-angiotensin-aldosterone system tends to increase blood pressure, to cause salt retention, and to promote fibrosis (scarring) and thrombosis (clotting) while the kallikrein-kinin system tends to reduce blood pressure, to cause salt excretion, and to promote fibrinolysis (clot dissolution).

I have been involved in patient-oriented research through the Vanderbilt General Clinical Research Center (GCRC), first as a clinical associate physician and then as an associate director. I currently serve on the National Institutes of Health (NIH) study section that reviews GCRC programs nationally. My major institutional responsibilities focus on the training and career development of clinical investigators. I direct the Vanderbilt Clinical Research Scholars program, one of sixteen NIH-funded institution-wide programs in the country. This program provides salary and research support for junior faculty dedicated to careers in clinical investigation and their mentors. I also codirect the Vanderbilt Master of Science in Clinical Investigation program, a two-year degree-granting program that provides mentored research experience,

Nancy Brown and her family
boating on the Rhein on sabbatical

3

coursework, and career development programming for postdoctoral trainees in clinical investigation.

I grew up as an air force brat, the youngest of three children. We moved eight times before I left for college, but my parents worked hard to provide stability in our family life and to portray moving as a great adventure. During my younger years, we lived in town, rather than on an air force base, and we attended local parochial schools; however, as my father's career required him to live on base, we attended military-subsidized schools and the quality of our education was somewhat uneven. My parents were the first in each of their families to obtain a college degree and valued education highly. Their values did not always coincide with those of the local school board. I never had a specific revelation that I wanted to become a doctor. The only physicians to whom we were exposed as children were military physicians, usually in smoke-filled, walk-in clinics. I do recall reading a series of nursing books around the age of seven and declaring to my parents that I wanted to be a mother and a nurse when I grew up; my parents suggested that perhaps I should consider being a mother and a doctor.

By the time I left home for college, dinner conversations had prepared me for a career in organizational management more than for a career in science. At Yale, my academic interests were initially undifferentiated. I took a lot of English courses and dabbled in the literary scene. But during sophomore year, I was swept away by Biology 120. The course was taught by articulate Nobel laureates who were passionate about their work. Hearing them describe how they had worked out the lipid bilayer structure of the cell membrane or other features of cell biology was far more engaging than the botany and phylogeny I had associated with high school biology. I decided to major in molecular biology and biochemistry, partly because of this focus, and partly for a more practical reason. I was on the women's crew, we practiced in the afternoons, and molecular biology had the fewest afternoon laboratory requirements of any of the science majors.

My first research experience came in the summer between junior and senior years of college. My parents were stationed outside of St. Louis, and I spent the summer working in the laboratory of Oliver Lowry, then professor emeritus of Pharmacology at Washington

University. Phil Needleman was the chair of Pharmacology and was studying chemicals called prostaglandins; I had no idea what prostaglandins were but the graduate student with whom I worked assured me they were "big." Lowry was an inspiring man. However, I found the work frustrating. To avoid evaporation, we were conducting biochemical measurements in microliter volumes under oil in a machined predecessor to the 96-well plate. It was before automated pipettes, and it took me most of the summer to obtain an accurate standard curve. I vowed never again to do research and returned to college. At Yale, Ethan Nadel, a physiologist at the Pierce Foundation and fellow in my residential college, convinced me that if biochemistry was not my thing, perhaps human physiology was. We undertook a study of the effects of heat on the time to reach anaerobic threshold. I had broken my leg and could not participate as a subject, but I enlisted many of my rowing teammates and my future husband. I vaguely recall obtaining informed consent, but I feel certain that at least one of the subjects met the criteria for a vulnerable subject!

I left college not at all certain that I wanted to enter a career in medicine or science. Ethan Nadel convinced me to take the MCAT exam, anyway. I worked for a year at a start-up energy company during the oil crisis of the early 1980s, for the shockingly large salary of $21,000 per year. We wrote many proposals in response to government requests for applications. These were typed on a bank of word processors that filled a room. After a few months of this, I decided that a business career was not for me and applied to medical school.

In 1982, I matriculated into the Health Sciences and Technology program, a joint Harvard/MIT program that included twenty-five students who took a unique core curriculum and completed a research thesis. I worked on an insulin nasal spray in the laboratory of Alan Moses and Jeff Flier. To my dismay, I learned more about the physical chemistry of bile acids, which we used to solubilize insulin, than about the bioavailability or pharmacokinetics of insulin in humans. Nevertheless, the laboratory became my home during medical school and offered my first real exposure to physician-scientists. When it came time to select a specialty and residency, my decisions were governed by other life choices. I had loved my surgery rotation, but by that point I was married and thinking about having a family; the few women surgeons I knew were divorced. So in the end, I decided to go into medicine. In addition, my husband and I decided that, if we were to manage two careers and a family, we should move back to Nashville, where his parents lived (mine were still mobile). And so it was that I began my internship at Vanderbilt in 1986.

John Oates, then chair of Medicine, inspired me and nudged me to pursue a career in Clinical Pharmacology after residency. I began my fellowship in the laboratory of Bob Branch, studying the effects of caffeine on renin release. Halfway into my fellowship, Branch left for Pittsburgh and our division director, Garrett FitzGerald, left for Ireland. However, by this point, I had found an area of research that really intrigued me, and I had tasted some success. So I continued with several studies involving the effects of angiotensin II and bradykinin with John Nadeau prior to completing a chief residency year. When I returned to the faculty, I was able to parlay these studies into mechanistic studies of the differences between angiotensin-converting enzyme inhibitors and the newly approved angiotensin receptor blockers. I learned the importance of mentorship the hard way during the next three or four years. Without a primary mentor, I made several potentially disastrous mistakes, including waiting too long to submit a FIRST award (a specific award for new researchers to help them establish their own research programs) from the NIH, resulting in the lapse of my funding for a year, and writing an R01 (a grant) on which I was not the principal investigator. That I survived in academia may be credited to the environment of Vanderbilt and the Division of Clinical Pharmacology, which combined scientific rigor with collegiality.

My husband and I had met at Yale. After graduation, Andy joined the marines, and we commuted for three years between Cambridge and either Quantico or Camp LeJeune. We married during January of my first year of medical school; in May, Andy deployed to Beirut. It would be impossible to describe the disconnection between the relatively self-indulgent concerns of medical school and the most primitive concerns regarding survival and communication. In those pre–cell phone, pre-Internet days, mail delivery could lag three weeks or more, and it took a week after the headquarters building was bombed in October of 1983 to learn whether Andy was alive. Andy left the marines in 1984 to attend Harvard Business School, and we graduated together in 1986.

We had our first child in March of the third year of my residency. I took five weeks of maternity leave (not enough) and had arranged my schedule to have only consult months left when I returned. I was nursing, and on the few nights I had to take call in the emergency room, Andy would bring Dan to the hospital for his late-night feeding. I delivered our second son, Ike, in August shortly after joining the faculty and wrote my Clinical Associate Physician Award during that maternity leave. The first month that I took call as an attending physician after returning to work seemed physically impossible. Lee Limbird, then chair of Pharmacology, wrote me a two-page, single-spaced letter about juggling family and career; I will always be grateful for that gesture of support from a professional whom I admired. Our third son, Sam, was born in the summer of 1996.

Our boys are now sixteen, twelve, and eight, with varied interests and passions. They attend University School so I drop them off at school on my way to work and can easily jog across the street to attend school events. The older two boys walk across the street to my office after their afternoon activities. I typically spend from 8 A.M. to 6 P.M. at work and tend to eat lunch at my desk. I schedule several large blocks of time during the week for writing. Otherwise, meetings would be all-consuming. I spend a half-day each week in the hypertension clinic, and I realized several years ago that this could only work if I did not serve as the primary care provider for my patients. When our children were younger, I tended not to get much work done at home because we kept them up late and I was usually exhausted by the time they went to bed. Now that they have homework, we often sit and work together in the evenings. Travel wreaks havoc with our carefully balanced schedules but cannot always be avoided. Weekends include sporting events. We are also very involved in Congregation Micah, our reform synagogue. I have chaired the religious school committee and teach third-grade Hebrew weekly. Judaism informs much of how I see the world, including the importance of time for rest and reflection, for study, and for family.

Juggling a career with raising children, tending to a marriage, and nurturing friendships is not easy and requires a sense of humor. At those fleeting moments when I have felt that I have had the balance just right, I felt that everything got almost enough attention. Still, an illness, a childhood trauma, a loss can quickly throw this delicate balance into catastrophe and serve to remind me of my physical or emotional limitations. It is in these moments that friends and family are most important. Intimate friendships with my sister and a few other women have been sustaining. Nevertheless, it is worth remembering that there are many women who juggle work, family, and friends with fewer resources than I, or anyone in the medical profession, is fortunate enough to have. For example, from the time our first son was born, my husband and I recognized that, for us, the advantages of having our children at home with a nanny outweighed the expense. Similarly, we could afford to outsource chores such as household cleaning and yard work so as to maximize our time with our children. And, while our decision to stay in Nashville when our children were young may have compromised our ability to negotiate for the most competitive job offers, our boys have benefited enormously from growing up in the arms of extended family.

I write these reflections just two months after Larry Summers's infamous remarks at Harvard regarding women's representation in tenured positions in science and engineering at top universities. That Summers misunderstands and incorrectly prioritizes the barriers to the advancement of women in science is evident to any woman in the field. That many men in science share Summers's viewpoint is unfortunately also an evident reality. It takes time to change a culture. Nevertheless, there are signs of hope, and the number of enlightened male leaders promises to increase as more men are married to, parent, or are raised by professional women.

Despite the challenges, I feel fortunate to work in medical research. There are very few careers in which one can see so readily the applicability of one's work. As much as writing papers and grants leaves one vulnerable to criticism and rejection, the process also serves to catalyze new ideas. Collaborations can lead to tremendous creativity, as well as lifelong friendships. Trainees continually invigorate the laboratory with their fresh perspective. The ability to engage in meaningful work is a gift I do not take for granted.

Chapter 2:
TINA V. HARTERT, MD, MPH
Assistant Professor of Medicine

I never imagined that this is what I would be doing with my life. I don't mean being a physician or my personal life, as I always imagined I would probably have a career, get married, and have a family. I mean being a scientist. Even in medical school, I had little idea of what it meant to be in academic medicine. In fact, I went to medical school to become a clinician and had only heard rumors about research. I actually wondered why sane, intelligent people chose to become scientists in the first place—what I heard about research was staying up all night to meet grant deadlines. I had absolutely no real sense of what it meant to be an investigator. What I am doing now is the direct result of trying things out, testing new waters. What seems simple on the surface is instead the result of years of figuring out what I wanted to do. As far as I can tell, I wasn't programmed for this career from birth. While I have worked hard, I have learned that this unplanned pursuit has opened the door to my soul. What follows is how I choose to answer the question of how I got to wherever people think I am today.

I am an assistant professor of Medicine and a physician-scientist. My research group's broad interest is asthma epidemiology. We study environmental influences that impact asthma development and disease expression. I spend about 80 percent of my time focused on my research, and about 20 percent of my time doing clinical work—either attending in the intensive care unit or in outpatient pulmonary clinic, or administrating programs. I

*Tina Hartert and her family
on vacation*

head—although I would rather say collaborate with—a research group of nearly twenty people including postdoctoral fellows (PhDs and MDs), biostatisticians, computer programmers, database managers, research nurses, and an administrative assistant. In addition, I have the privilege of collaborating with people throughout Vanderbilt Medical Center and University, and several other medical schools throughout the country. My work allows me the opportunity to learn and share with scientists all over the world. This sharing of ideas and creative energy is tremendous (big emphasis) fun.

I was born and spent my early years in England, the oldest of three children. My mother is British, and my father, now deceased, was the child of an immigrant from Luxembourg who came to this country via Ellis Island. I had no exposure to medicine or science growing up. My mother had studied foreign languages and my father journalism. My mother is a very organized and capable woman—a quiet saint, unflappable, a doer who at an early stage showed me that women are able to do what they set their minds to do. If we needed something, she would simply make it. If a teacup needed to be repaired, she did so rather than buy a new one. When our community needed a pool, my parents organized the community, raised money, and got a pool built. Few families where I grew up had money for college, so my parents organized and solicited businesses and started a scholarship fund for our high school so that college could be an option for more students. It was important to my parents to give to the community—to build community—and they never sought recognition for what they did (my mother's definition of a true leader). My parents always talked about the importance of giving back in gratitude for all we have been given in life, and they served as surrogate parents to numerous other children. My father less quietly encouraged me; in fact, he was less quiet about most things. Although he had no scientific background whatsoever, except a very good friend who was a physicist, he decided that being in science or medicine would "suit" my academic talents and interests. He was a very uninhibited person and honestly didn't care what anyone thought about what he said or did. He was a thinker, a builder.

I earned my bachelor's degree with dual majors in human biology and European literature and society from Brown University in 1985, my medical degree from Vanderbilt University in 1990, and my master's degree in public health in 1998. After completing an internal medicine residency and chief residency at Johns Hopkins in 1995, I did my pulmonary and critical care fellowship training at Vanderbilt and joined the faculty in 1999. While I was a resident, I married Stokes Peebles, also a physician-scientist, and we have three children. Although my parents tried to instill in me the belief that I could do anything that I wanted to do, I haven't always believed it—my children, for one, have certainly taught me otherwise.

Most people go into medicine because they want to care for patients, as was the case for me. Most of them also have no idea what it is like to be a scientist, as was also the case for me. Medical training doesn't offer much glimpse into the life of a researcher. Even today, when I interact with the average medical student, it is most likely to be in the clinical arena. For me, I entered research as if I were entering uncharted waters, with a tentative dog paddle at first, all the while thinking, "I'm not sure this is for me." My primary research exposure began while I was a medicine house officer at Johns Hopkins, when a group of fellow house officers and I came up with the idea of using oral antibiotics to treat the enormous number of febrile intravenous drug users on our service with right-sided endocarditis (infection of the heart). The benefit would be to provide four weeks of therapy at home without the need for intravenous access. Some thought we were crazy to treat systemic bacteremia with oral antibiotics, but it worked. Next I studied asthma, because I had really become hooked on pulmonary medicine and the clinical aspects of the intensive care unit. Still, I had little idea of how an average researcher spent her days, weeks, and years outside of clinical practice. I didn't know what an attending meant when he or she went "back to the lab." I believed I was a pretty good doctor, and I liked it, so why would I forge into uncharted territory?

Following residency, I spent two years in purely clinical work and teaching in an academic setting as a clinical faculty member. There is no doubt that being a clinician is hard work and a labor of love, with long hours, lots of stress, never-ending paperwork, and sometimes poor outcomes because of biologic variables you cannot control. There are very positive sides as well—the healed patient, as well as patients and families with whom you go through the cycles of life. However, I found it emotionally draining, as if the work used me up and spit me out again. I have incredible admiration for those who practice clinical medicine full-time; however, I find being a physician-scientist a refreshing balance. I can maintain a love for clinical practice and the energy it requires precisely because I'm not doing it all the time. What drives my commitment to science most strongly, however, is the realization that medical progress is impossible without research advances and the committed physicians to apply that research. The ability to see your science make an impact and be put into action is extremely rewarding.

When I transitioned into fellowship, I certainly floundered at first. I did something that I now tell graduate students and fellows never to do. I decided to develop a research area on my own as a postdoctoral fellow, rather than work on someone else's projects. I did this because no one was doing what I wanted to do. This was a mistake. Working within a group, with a mentor, and on projects with a funded infrastructure, is the best way to train. In this way, you develop your skills by working alongside an established investigator who knows what he or she is doing and who can then help you find your own niche. I am incredibly lucky to have made it this far having taken the other route.

One of the most important jobs that I do is to show trainees that medicine and science can be fun and rewarding. I also try to show them the full spectrum and diversity of a career in academic medicine. As a mentor, I receive incredible pleasure from seeing someone with whom I work succeed. I am passionate about encouraging students, residents, and fellows to enter academic medicine, not only because it can be a fulfilling career, but also because it offers a way to make a lasting and positive impact—on students, on the field of science, and on the lives of countless patients. By improving medical knowledge and patient care through research, an individual's impact can be greater than by only treating individual patients.

Mentors are incredibly important. They shape your view not only of academia and your career, but also of the world. One of my mentors, Marie Griffin, once told me, "I need to feel like what I do makes a difference." This is simple advice that anyone can understand; I've taken it to heart so that what I do for a living is "what I do," not work. Griffin taught—and retaught—me the basics of scientific inquiry; she still teaches me these truths, although I've never heard her call herself "a scientist." Another mentor, Bob Dittus, showed me the joy of building and creating. He set a precedent on balancing work and family; he worked hard but he would also be the first person to say he wouldn't be somewhere because seeing one of his daughters play tennis was the most important item on his agenda that day. These professional and personal friendships sustain me and have helped my career.

Building a community through networking and mentoring is vital to the ongoing success of medical research. It is for this reason that I am much better suited to an academic research career than clinical medicine, but that aspect of academics wasn't clear to me early on. Collaboration is one of the features that I enjoy most about academic medicine, and it comes easily for me. Colleagues fill my gaps and do what I cannot do, and I fill theirs. The whole is greater than the pieces. Collaboration and mentoring are about the desire to engage others, to work together, and to share. They are about humility of attitude and transparency of motivation, and a sense that all these attributes should be common to humankind, not just science, because collaboration values sharing and decency. So, thanks to what I learned from Griffin— her skill and her humility—I will always be giving back, sifting through the days ahead, keeping an eye open for all the good stuff to still be discovered, and helping others discover it as well.

I tend to see the glass half full, rather than half empty, and that probably influences my perspective on being a woman in science. Although I am relatively young, I was actually the first female faculty member to take maternity leave in our division, which was composed of mostly men. My division chief was extraordinarily supportive during my maternity leaves and made this a pleasant and unpressured time for me. Supportive colleagues and a supportive environment are tremendously important. I have never really felt any overt discrimination within my division; although when we were first recruited to Vanderbilt, I didn't get a separate offer letter from the dean—mine was a paragraph at the end of my husband's letter! I've also heard plenty of comments that I care not to repeat, and have certainly experienced things that I often think would never happen to or be said to male colleagues, but this probably happens more often outside of work than at work. All in all, I work in a great environment with fabulous colleagues who inspire me and are a constant source of creative energy.

The first meeting of a group of women scientists, where the idea for these essays was conceived, was an eye-opening moment for me as I recognized truly how few women physician-scientists exist. As I approached the room for the luncheon convened by our department chairman, I thought I must have the wrong location. There are 460 faculty members in the Department of Medicine, and this was such a small conference room, it could not have been correct. It was correct—there were about twenty women in the room, and the number of senior faculty could be counted on one hand. I don't feel lonely in this work, but I do think women in training are often more naturally drawn to or seek out a female mentor. We need more of us.

Balancing work with family life is no doubt a challenge, but I know that being a researcher makes me a better person and mother. I have a need for intellectual challenge, to think, and to be creative outside the realms of motherhood. Fulfilling these needs makes me more content. Our children see that I enjoy what I do, that I make a difference, and that I am not one of the parents who *have* to go to work each day. To effectively accomplish all this, setting priorities is very important. I know what the most important things are in my life. Those tasks that have to be done but are not priorities, just part of everyday living, I divest to the people around me that make up my support system—a housekeeper, a yard keeper, an administrative assistant, etc. Additionally, I have a wonderful nanny and chauffeur to run the show while I am at what I hesitate to call work. It's not that I want my children to see that you can pay someone to do everything, but I do want them to know that I would rather get help from others in performing certain daily tasks so that I can focus on the matters that are most important to me. I do take on the primary responsibility, however, for running our home, making certain that we have quality child care, planning meals, arranging our children's schedules, helping with homework, and so on. I think that often falls to women, even in dual-career marriages, although I mostly enjoy these duties. Therefore,

I think there isn't much buffer room for women; if something goes wrong in "the system" it is hard to keep up the same pace. That's when being a member of a larger research group of supportive colleagues is critical. Additionally, for women who do have families, maternity leaves generally occur during a very formative period, during early developmental faculty years. I have taken three maternity leaves during my junior faculty years. However, I realize that children grow up all too quickly, and I am not willing to miss out on what has been God's most wonderful gift to me. Academic medicine allows me a flexibility that I wouldn't have in clinical medicine.

How does it work for me? I am a night owl, and it is a time when I am most productive; I love the peace of the evening and uninterrupted time to think and write. I'd say that I am pretty organized and efficient—my husband describes me as a "perpetual motion machine." I like to be building and creating things—I definitely thrive on that. So, here I sit in my peaceful kitchen, as I do nearly every night, working and writing, knowing our children are peacefully sleeping as our dogs sleep next to my feet. I started the day together with my family, took my children to school, was home by six o'clock and spent the entire evening with my family, not doing the busy things of life, but really spending time together. When I am not on call, I also have the flexibility to leave work to see my children play soccer or swim—a priority I set for myself and for those in my research group.

Achieving balance is difficult and I don't pretend to have all the answers, but I've found it easier if I set priorities. Life goes by too quickly—I would say, pick what you enjoy and spend your time doing it. The things that are most important to me have no monetary value—even in my work, it's the emotional income that keeps me in academic medicine.

Chapter 3:
LEE E. LIMBIRD, PHD
Professor of Pharmacology

I am currently vice president for Research at Meharry Medical College and chair of the multidisciplinary Department of Biomedical Sciences. From 1979 to 2004, I was a member of the faculty of the Department of Pharmacology at Vanderbilt, serving as departmental chair—the first female chair in the School of Medicine—from 1981 to 1998. From 1998 to 2003, I also served as the first associate vice-chancellor for Research at Vanderbilt University Medical Center. My laboratory colleagues and I have received many national awards for our research findings concerning α_2-adrenergic receptors, their trafficking to various cell surface compartments, and coupling to multiple cell signaling pathways. I have had the pleasure of chairing national meetings and national advisory committees for the National Institutes of Health (NIH) and private foundations. I edited *Goodman and Gilman's Pharmacological Basis of Therapeutics,* and I authored my own textbook, *Cell Surface Receptors.*

To my parents, a college education was everything. It was a privilege they didn't enjoy, so what they wanted for my older brother and me was nothing more or less than a college education. In fact, we were told that we were "planned" five years apart so it would be financially more feasible to educate both of us. Our family was not financially fortunate, but in every other way we knew

Lee Limbird and her lab staff pop a cork to celebrate an accepted paper.

abundance, particularly in terms of optimism about our limitless opportunities. My dad would often tell me that I could "be anything I wanted to be," and he really meant it. My mother's style was different, but her expectations of excellence were very, very clear.

Frankly, I don't think I fit the profile of someone who had clear career goals or aspirations when I was young. I was a bit chatty in high school and often found myself in the corridor for having too much to say in an overcrowded classroom. As a result, I didn't get into my first choice college, something that turned out to be quite fortunate. My jobs during high school were simply to raise money, with no attempt to seek professional experience that might be relevant to a later career. I served coffee on an early morning commuter train, was a lifeguard at a local YMCA, and later in high school, would spend my summers with my aunt and uncle in the Pocono Mountains of Pennsylvania so that I could waitress at the Split Rock Lodge resort and at a nearby restaurant. One of the gifts of those work experiences was that I came to learn how hard many people struggle to put together enough income just to get by. These workers, of course, were the same taxpayers who later would be funding my research. This realization remains humbling.

Although I didn't decide for sure where I was going to college until after I graduated from high school, I matriculated three months later at the College of Wooster, a small liberal arts college in Ohio with one unique feature at the time: an independent study program. The fourth year for every Wooster student was spent working with a single mentor on an independent study project. During my independent study project, I experienced a bit of liberation from my fear of failure and began to learn how to learn on my own, something that would bring me great pleasure and serve me well professionally for the rest of my life. These two mantras of the College of Wooster—freedom from the fear of failure and lifelong self-learning—were so empowering to me that I later incorporated them into expectations of our PhD training program in Pharmacological Sciences at Vanderbilt.

My interest in science probably began with my older brother, Jay, who ultimately became a chemical engineer after graduating from the School of Engineering at the University of Pennsylvania. I looked to him for guidance on how I might live my life, although I doubt I was aware of this at the time. I took science in high school, but I was not truly intellectually engaged—the laboratory courses were monotonous and the outcome was known. During a high school career day, I learned that medical technologists do the same experiments, or assays, over and over again. So medical technology, a career I had considered, no longer sounded like a match for me.

I was seduced by science at the College of Wooster. Faculty in the Department of Chemistry took a particular interest in students and had a contagious enthusiasm about the particular area of chemistry with which they were involved. A number of Wooster faculty and staff were particularly valuable to my career development; my chemistry mentor for my independent study, who has become my lifelong mentor, was Theodore Roosevelt Williams, or "TR" to me. TR is African American, and I learned quite a bit from him about navigating obstacles, certainly insights that served me well in biomedical research. I found his courses on analytical chemistry fun—sleuthing the identity of chemical compounds really caught my attention. My independent study at Wooster was a clinically relevant analytical chemistry project. Frankly, the clinical relevance served two purposes—it was something I was interested in and a way for me to visit Duke University, where my fiancé, Tom Limbird, was in medical school. My co-mentor for my independent study project was Galen Wagner, who directed the Coronary Care Research Unit at Duke Medical Center at that time.

Graduate school might have been the obvious next step for some, but I took that step for nonobvious reasons. As the wife of a medical student at Duke University, job opportunities were slim, as folks in Durham found the probable short-term nature of those hirings undesirable. After all, your husband would graduate in a few years, and there was always the possibility you would get pregnant; it was quite a different time, then.

I applied to graduate school at both Duke and the University of North Carolina and was fortunate to be accepted to both. I chose Carolina for my PhD over Duke because the faculty seemed less arrogant; I was definitely a product of the '60s!

After nine months in the biochemistry program, I left graduate school. I had taken all the courses for my PhD in that short time, as we were living on my graduate student's stipend, Tom was always in the hospital, and the only thing I could do for free was take courses! However, the faculty members I associated with at that time were largely unenthusiastic about their teaching and seemed even less enthusiastic about their research. I was concerned that I might have entered a graduate training program where I was not interested in the career it made possible. Leaving graduate school temporarily was impulsive, but it was the right decision, as it turns out. With Galen Wagner's help, I was fortunate to find a research position at Duke University that followed up on my Wooster independent study project. I developed a method for quantifying the isoenzymes of CPK, which later became a routinely performed diagnostic assay for coronary artery disease. This work ultimately became the dissertation research for my PhD. I jokingly tell people that I got my PhD in three years because I didn't like graduate school, so I really didn't want to stay long; but there is great truth in that intentionally humorous statement.

A number of advisors created possibilities for me in my scientific career. For example, Andrew G. Wallace, chief of Cardiology at Duke University, was willing to fund me as a cardiology fellow for my postdoctoral training in part because, in his mind, my graduate work had considerable clinical significance; reflective of those times, I ultimately was not permitted to first-author any of it. Andy Wallace introduced me to Robert J. Lefkowitz, who was just joining the faculty at Duke University. This was a great stroke of fortune for me. Secondly, James Wyngaarden, then chair of the Department of Internal Medicine, was philosophically enamored with training PhDs in clinical environments and, reciprocally, placing clinically trained individuals in basic science environments for their fellowships. He believed, and I think he was right, that these training experiences would foster what we now refer to as bench-to-bedside research.

One of the blessings of being a Duke cardiology fellow was Saturday Grand Rounds, which started at 7 A.M. and ended when they ended. Eugene (Gene) Stead, an intellectual force in academic medicine throughout his career and a legend in his own time at Duke, was a predictable attendee. Gene Stead's mantra was that he learned things not when he read them and not when he heard them, but when he turned around and taught them to someone else. I have found this insight to be true throughout my career.

As Bob Lefkowitz's first postdoctoral fellow, I had the privilege of watching the laboratory grow from two people to twenty-two, the size it was when I left Duke University six years later. In the early years in Bob's laboratory, he would end most days by taking a seat on a stepping stool in the corner of the laboratory and telling stories about the people in science: what motivated them; how their personal traits influenced their selection of a biological problem to solve or the styles with which they solved them; how their challenges had made them more fierce in achieving their goals; and how people found myriad strategies to ultimately succeed in their career. Many of these stories were laced with terrific humor, as Bob is an extraordinary raconteur. What I learned from these stories was that people in science are very diverse and extremely human, and that there is richness in their science beyond the data. These insights were powerful in my understanding—ultimately in my own laboratory—that each person is an individual with individual gifts and traits. Helping my trainees find their strengths maximizes their enjoyment of science and also their contributions to it.

Bob is more than a raconteur, however. He also emphasized the power of data. Data are the currency in science. There was a trainee, Nanette Bishopric, in the early years in his laboratory, who once wrote an "Ode to Bobby Jo" (our nickname for Bob—another long story), and the refrain in each of the stanzas was "bring me the half that's got the data." Bob's emphasis on data and what the data say was a powerful mentoring message.

Fortunately for me, the appearance of mentors in my life did not end with my postdoctoral training. When I joined the faculty in Pharmacology at Vanderbilt, I knew virtually nothing about the formal discipline of pharmacology. My degree was in biochemistry, my postdoctoral fellowship was in cardiology, and my postdoctoral laboratory training focused on simply answering questions about the β-adrenergic receptor. Although β-receptors bind ligands and regulate signaling pathways, both paradigmatic concepts in pharmacology, my laboratory studies of the β-receptor had not resulted in my becoming a scholar in pharmacological principles. Fortunately, my chair, Joel Hardman, was fully aware of my knowledge base and its limitations. I had the blessing of his mentorship in myriad areas. Interestingly, coming to understand the discipline of pharmacology informed my research in ways I might not have anticipated in the arrogance of my youth.

There have been other mentors, many of them my national colleagues who have been both brutally honest and at other times reassuring listeners and advocates. For example, I knew if I called Al Gilman (at the University of Virginia and subsequently at the University of Texas, Southwestern), I would get the "honest truth," whether I wanted it or not. I also had the blessing of wonderful collaborators over the years, some chronological peers, some younger, and some senior to me. In all of these circumstances, I learned more than I contributed and looking back, my national and international colleagues contributed significantly to the fabric of mentorship. Brian Kobilka, for example, has been a source of innumerable insights, and I continue to admire both his mind and his professional style.

Early in my career, I was so focused on getting answers and writing results that I fear the students in my lab were overmanaged. Nonetheless, they also served as teachers and mentors of me, their advisor. I remember very clearly one day in our weekly lab data club when David Sweatt, then a graduate student and now chair of Neuroscience at the University of Alabama, challenged me by saying, "What is your reason for purifying the α_2-receptor, just to get a single band on a gel? Is that all you want? What do you want to *do* with the purified protein?" I had allowed myself to get so immersed in the details of purifying a trace membrane protein to homogeneity that I had lost sight of the "so what?" What a blessing that I had my trainees to keep me honest! I fostered a challenging and intellectually aggressive environment in the laboratory; the product of my efforts was colleagues who challenged me as well, and I was gratefully honed in the process.

My understanding is that one of the reasons for compiling this volume of stories about women, their lives, and their careers in science is to get a glimpse of how these women "did it all," which, by the way, I think we would all agree that we didn't! My perception is that it is when scientists start their families that the disparities between male and female professionals begin, and I don't think this parenting-provoked disparity occurs only in biomedical research. I once read, and I will poorly paraphrase it here, an opinion of Margaret Mead: Women can't decide about their ultimate career goal until they decide whether or not they will have a family, and when. This was certainly a true statement for me. In our twenties, my husband and I learned that we would not have our own children. Without children as part of our future, I considered going to medical school.

However, for a variety of reasons, I could not get individuals around me to enthusiastically endorse my medical school matriculation. There is no doubt, however, that learning we would not have children caused me to begin to think for the first time in a serious way about my career, and to make very intentional choices. I began to invest intellectually in my research questions in a way that I had not done previously. I became extremely critical about the questions that I asked and their importance and extremely focused on advancing the field and its understanding, rather than just being busy. Ultimately, Tom and I decided that not having our own children didn't mean that we couldn't have a family, and years later we adopted. We adopted our son only to learn that he had a sister, and so we adopted her, too, and our family grew to include Eric and Jessica.

In our case, adopting children when we were thirty-six and thirty-seven meant that the impact of parenting on our career trajectories was not all that significant. For example, I already had tenure. In fact, having a family, ironically, was a professional boon, as my commitment to our family caused me to take much more responsibility for time management and brought a focus to my research program that I hadn't taken when I felt I had unlimited professional time.

Parenting, however, did cause both Tom and me to reevaluate our priorities. It was clear that when we were at home, we wanted to be engaged with our children, teaching them a view of life that we thought was important to share. To protect our family time, we paid individuals to run errands for us, clean our house, and perform a lot of other repetitive tasks of living. We didn't have live-in or full-time help, and that, too, was an intentional choice. For example, we chose to make our own meals together, which was really quite a bit of fun. We never confused taking care of our home with taking care of our children. I think it's important that the babysitters we had for our children after school, most of them local college students whom we recruited by paying them more than they could earn doing work-study, never had household tasks to distract them, and our kids fully understood that they were much more important to us than taking care of our home.

It is premature to assess whether we made good choices in our parenting, our parenting styles, and how we chose to balance our many lives. But our children already have gifted us with gratitude for many of these choices and our efforts. If, when our kids are fifty years old, they can look back at their lives and feel they contributed to society in positive ways, with compassion and gentleness toward others, then Tom and I will be able to say we achieved a measure of success in parenting.

I do have some humorous stories of what happens when one raises a family at the same time that her mind is focused on answering biological questions in the laboratory. Our son, Eric, for example, likes to entertain everyone with stories of the times I would come to soccer games and cheer whenever a goal was made, whether it was his team's or not, because my mind had wandered off to some other place. Our daughter, Jessica, has similar stories, like the day I drove her to the airport instead of her piano lesson because I was so frequently going off for one-day trips that my car seemed to find its way to the airport on autopilot.

People mention the word "balance." I don't exactly know what balance means. If it means taking a pie and cutting it into eight pieces and making sure that your life has one-eighth of each slice of life, I wouldn't call this balance. I would call this an insipid way of living. In contrast, for me balance can better be pictured like a bull's-eye. In the center is what's most important to you—your spiritual life, your family, whatever you choose. Circulating out from that center are things of decreasing importance. If you know what's most important in your life, then when crises occur, you also know where to focus your efforts, and what to ignore. Contemporary writings about balance irritate me more than they satisfy me.

My advice to young scientists, both male and female, can be divided into the philosophical and the practical. Philosophically, I think that scientists must remember that it is a privilege to be in this profession. How many people can get up in the morning, put their feet on the floor, and know that they have the privilege of following their own curiosity? With this privilege, however, comes responsibility. We use public funds to support our research and the concomitant training programs in our laboratories. Consequently, it's important in our research to focus on what we think is the most critical scientific goal and thus advance the field in significant ways. It is not appropriate to spend your energies looking over your shoulder to figure out what competitors in the field are trying to do and simply beat them. If competitors are doing it, let them do it. Choose something that advances the field beyond what your competitors may have the courage to do and, as a result, gift society with the new knowledge that it needs. Finally, I also think it is a gift to have a profession that has no time limits on it, a job with no time clock. A lot of verbal time is spent discussing the hours that one spends in research. Frankly, I grieve for those who have a forty-hour-a-week job they can leave. My dad supported our family throughout his life by sorting mail for the U.S. Post Office. For a variety of reasons, despite his considerable intellectual gifts, he was fortunate to have this job, but his life truly began when he left work. I would want for all people the blessing of a job that so engages their passions that they feel both intellectually as well as personally gratified by it. That is how I have found biomedical research.

Practically speaking, data are the currency—if you focus on data, then progress and occasional discoveries will be the consequence. Focusing on getting ahead, whatever that means, is a distraction and usually leads to forgetting that new findings are the currency, and the higher the impact of the findings, the higher the denomination of the currency.

For women and minorities, remember that you will have your genotype your entire career. I once arrived at a Cold Spring Harbor meeting apparently looking terribly exhausted and drained. The late Eva

Neer, a Harvard professor and member of the National Academy of Sciences, told me that I looked like I was doing too much, a euphemism for "you look terrible!" When we went through the list of my commitments, she reminded me that I would always be a female. Yes, she explained, females will always be needed on national committees; females will always be needed on local committees. Basically, during the run of my career, I would be able to contribute in all ways. However, her advice was that I simply could not do it all at any one time. I needed to choose where I would put my efforts during different seasons of my life. This was probably the best-timed and most powerful advice that I ever got from any one person in my career.

Focus. Selfishness with your time at the beginning of your career will result in your being able to contribute in many dimensions later on. The most important accomplishment early in your career is to establish your credibility and capability of contributing in your field. Once this has been established, what you can do on behalf of your trainees and your institution is significantly enhanced.

Be rigorous about time management. Designate a time each day to interact with each new student and postdoctoral fellows in the laboratory. After they get their bearings, make sure you spend an hour or so with them at least once a week. Decide how many journal articles you will read and understand in detail each week, and do it. When you serve the profession, for example, by serving on an editorial board, decide what hours a week you will review articles. When that time is over, stop reviewing until next week. I watch so many people do first what is presented to them, appearing to wait until their desk is clean— as if this will ever be—before beginning their own intellectual work. Dive first into your own intellectual work and then manage the time of other professional commitments around that.

Be organized, knowing what little tasks and large tasks you have to get done. I always had a bit of work with me, a journal article to read or stationery to write to people I cared about so that I could make the best use of waiting room times for physicians, or in the mall while an adolescent shopped.

Always treat people like the special human beings that they are. There are many vagaries in science. There are days of success followed by months and years of quiet times until the next significant findings emerge from your laboratory. I have observed people who get inappropriately arrogant during their narrow windows of success and then find themselves terribly lonely in the long spells in between. Borrowing from one of the philosophies of the Shakers' founder, I think it is useful to live "as if you will live forever or die tomorrow." Keeping that in mind informs not only your decisions but also the way in which you interact with people.

Remember, research is a creative enterprise, not much different from that of artists, musicians, poets, and authors. You need to make time to muse. It is not enough to be busy in research; it is absolutely important to take the time to be sure you are busy doing the right thing!

Young scientists may find role models and inspiration for their scientific style from the lives of other creators. For me, the author Anne Tyler has provided some unwitting advice about manuscript writing. I read once that she develops the characters in her stories— she peoples her books with such delightful and unusual characters—by staring at framed photographs of different individuals until they take on personalities, and those personalities engage in relationships with one another, and then a story is born. From this, I learned to tape the figures for a new paper above my desk, or the desk of my trainees, and let the figures "speak." In doing so, the flow of the results section is established. From the results comes the flow of the discussion section, and from the summary comes what must be contained in the introduction section to appropriately introduce the reader to the content of the paper.

I also like Tyler's style of working when her children were young. After they went out the door to school, she went up to her study: no clean up, no loose ends, no phone calls. At lunchtime, she had a peanut butter and jelly sandwich, read the mail, and went back to her study until her children returned home. She knew how to manage her time so that she had time for all the things that were important to her.

It is hard to conclude this writing without expressing, once again, my gratitude for a life of unceasing blessings, including those of my professional life. I wish these privileges existed for all people; perhaps, in some way, each of us contributing to this volume can do our best to share the joy for our work with all whom we come in contact, and at least some of the financial benefits of our efforts with those in need. We can pass on opportunities wherever we find ourselves, in much the same way our families and mentors freely passed on opportunities to us.

Chapter 4:
MARIE R. GRIFFIN, MD, MPH
Professor of Preventive Medicine and Medicine

A tramp once knocked at the doctor's door,
And asked for the pants that the doctor wore.
A lady answered the door and said,
"I'm the doctor," and the tramp dropped dead.

As professor of Preventive Medicine and Medicine, I spend most of my time doing epidemiologic research primarily on drugs, vaccines, and vaccine-preventable diseases such as influenza. I am most drawn to research that has direct application to clinical practice and that can influence public policy. I currently see patients in Vanderbilt's internal medicine practice one day per week and now am reaping the rewards of having known some of my patients for almost twenty years. I am an attending physician at the Veteran's Administration Hospital one month per year, where the house staff helps keep me on my toes. For the past ten years I have taught epidemiology and public health practice in our masters of public health (MPH) program, and currently have four trainees going through that program.

As the second of seven children, with a straight-A older sister a year ahead of me and a walking-encyclopedia brother a year behind me, I was certainly not the stand-out student in our family. Although our parents were first-generation Irish Americans without college degrees, we all got the message that we would go to college—and become financially independent as soon thereafter as possible. Growing up in the '50s, that "Boom,

Marie Griffin hiking with her sons in the Pyrenees

Boom, Ain't It Great To Be Crazy" verse above appealed to me. Was a woman doctor really that odd? Is this a way I might inspire shock and awe? My high school classmates provoked me further—they thought I was joking when I said I was going to be a doctor. It was not something that girls or boys at my large parochial high school did. Fewer than half of our class went to college. I was the only one in my class of more than two hundred who went to medical school. I didn't believe my friends at the time when they insisted I was clueless and naive. Of course, they were right, but it was probably good not to know that then.

I was pretty good at science and math and loved doing the dissection parts of biology classes. I liked figuring things out, solving problems. I didn't really like school much, but a string of mind-numbing, low-paying summer jobs helped spur me on to medical school. Medical school at Georgetown was OK, but still school. It took internship to really engage me—the immediate response of blood, IV fluids, diuretics, or morphine; the stories; the comradeship; the heady power that accompanied the steep learning curve; the exhaustion that diminished other concerns; and, at last, financial independence. Still, at the end of three years of an internal medicine residency, I couldn't envision my life in medicine. A two-year stint as an epidemiology intelligence service officer at the Centers for Disease Control and Prevention (CDC) put into perspective the diseases that I had been treating up close. I was back on that steep learning curve again. Epidemiology took me back to my love of problem solving and gave me a framework and tools with which to work. So, I had two things I loved to do—medicine and epidemiology—and I was unwilling to abandon either of them for any length of time. I started to envision a life in medicine in which I might do epidemiologic research and still see patients part-time.

After the two years at CDC, I got my MPH at Johns Hopkins to help me better grasp appropriate study design for the observational research I hoped to do. About that time, I also got married, and my career moves were, shall I say, not completely self-directed. I joined my husband at the Mayo Clinic and conducted what turned out to be an invaluable two years of epidemiologic research. I had started seeing patients again at Hopkins, one day a week while I was doing my MPH, but I had relatively few clinical duties in the four years since my residency, so I signed on for a fourth year of general internal medicine fellowship. Basically, I was waiting for my husband to finish his fellowship and decide his next steps. However, this holding pattern turned out to be great preparation for a faculty position. I worked with experienced researchers, produced quite a few papers, and regained confidence in my clinical skills. Most momentous during what might be considered a prolonged fellowship was the birth of our first son. I took three months off—three long months in Minnesota with a very active baby, no family there most of the time, and most of our friends working. Here was another sharp learning curve. I thought it would be a breeze because of all my younger brothers and sisters and all the babysitting I had logged as a teenager. THIS WAS MUCH HARDER THAN INTERN-SHIP! Going back to work was a nice break.

I was recruited to Vanderbilt as part of a package deal as the trailing spouse. What was I looking for? I really have never had a five- or ten-year plan. I knew I wanted to combine research with clinical work. I wanted flexibility in my schedule because I knew I wanted more children and did not want to become overcommitted. I wanted to work with people I liked and who understood my priorities, and to work in areas I thought interesting and important. But there were and are lots of research questions I find interesting. I have never been driven by a particular question. Much of what engages me is trying to figure out the best type of study design to answer a particular question. In the Department of Preventive Medicine, I felt the freedom to set my own priorities with no pressure to travel the more common academic route, which often includes long work hours, frequent travel, saying "yes" to every opportunity to give a talk or write a chapter, or relentlessly pursuing a single research area in which one becomes the expert. My years of fellowship paid off, so that meeting the requirements for tenure was relatively easy even though I worked primarily eight to five when the children were small. I still don't have a five- or ten-year plan but continue to have a similar goal to the one I had twenty years ago. *How can I keep this fun and interesting and do research that could have an impact? How can I continue to work with people I like?*

These days, I drive into work with my fourteen-year-old son, the radio playing National Public Radio or hip-hop depending on whether he's alert enough to change my station that early in the morning. I drop him off at school across the street before I come to work. Throughout their school lives, our two sons have been within walking distance from my husband and me at Vanderbilt. When they were small, I took them to school most days and picked them up from after-school, or other, activities on my way home. I joined them for lunch (when they let me), coached their soccer teams, and devoted my weekends almost exclusively to them. My mother, who had seven children in eleven years and never worked outside the home after she had children, insists that I spent much more time with each of my children than she did with each of hers. I bless her for telling me that. Nowadays, with one in college and one who doesn't think he needs me at all, I do much more work at home. In fact, I'm looking for another excuse not to work all the time—but not ready for grandchildren yet.

Medicine offers many options for those who need or want flexibility in their schedules. I suppose I was naive about the demands of family and work, but many of those things are impossible to know or plan for ahead of time. Some children are easier than others, some spouses and families more helpful, and other circumstances are just unpredictable. Research has differing demands than clinical care, but to me they seemed to be more predictable demands. Choosing research early in my career did not close off the possibility of being a full-time clinician later, or using my training in some other way that fit in with my other priorities. Doing research makes the practice of medicine more interesting for me, helps keep me on that addictive learning curve, and I hope, makes me a better physician.

Chapter 5:
WONDER P. DRAKE, MD
Assistant Professor of Medicine

I am currently an assistant professor of Medicine in the Division of Infectious Diseases. I work as a physician-scientist, allocating 80 percent of my time to basic and clinical research on sarcoidosis (chronic inflammation of the lungs and kidneys) and 20 percent as a clinician for the inpatient HIV service. The work of my laboratory regarding the role of infectious agents, particularly mycobacteria, in the pathogenesis of sarcoidosis is receiving national attention. We have received invitations to present our findings at national meetings, as well as to the World Association of Sarcoidosis and Granulomatous Disorders. My laboratory has also been asked to participate in a national research program to investigate the role of infectious agents in sarcoidosis.

No one would have predicted I would be where I am today, but it is important to dream. I was born in a small town in Alabama, the third illegitimate child of a teenage mother. She married my dad, the father of her three children, approximately three months after I was born. They went on to have two more children, for a total of two boys and three girls. I am the middle child. We did not grow up in a white house with a picket fence but lived in a four-bedroom mobile home. Neither my mother nor my father graduated from high school. My mom was not allowed to be valedictorian of her class because of her pregnancy. She dropped out of school in anger. After his

Wonder Drake and her family vacationing

mother died, my father dropped out of school in order to take care of his seven siblings. While these are discouraging demographics, they do not relay the whole story. Both of my parents had incredible work ethics. They refused to accept welfare, although I am sure that there were times when we could have easily qualified for public assistance. I remember my mother saying, "I'll work anywhere to stay off of welfare." We were required to be respectful to any adult, to dress neatly even if we were going outside to play, to keep the house clean, and to do well in school. Report cards were serious business at our house. Everyone's grades were reviewed every semester, and unacceptable grades had consequences. For example, I remember my mother asking me why I had a B$^+$ in chemistry, even though that B$^+$ was the highest grade in the class. I must admit that I was envious of children who lived in nice houses and had educated parents. Now I understand that parents who love their children and are interested in their progress are high-quality parents. As a child, I had a dream of going to college and becoming a teacher or a physician. It really was a dream because no one in the generations before me had graduated from high school, and college seemed a world away. I worked really hard in school, received numerous academic (not one athletic) awards, and graduated valedictorian of my class.

I went to the University of Alabama, choosing microbiology as my major, with the goal of becoming a science teacher. However, all my college classmates and friends were pre-med majors. I started attending the pre-med meetings and, honestly, it just felt right. I knew deep inside that medicine was the right path for me. I hated the idea of not becoming a teacher, but I knew that I could also help society as a doctor. I performed well in college, became president of the Pre-Medical Society, and then came to Vanderbilt University School of Medicine. At the time, I envisioned doctors as working in private practice; the idea of academic medicine was foreign to me. The summer prior to starting medical school, I applied for a summer research position supported by Martin Blaser, who was chief of the Division of Infectious Diseases at that time. He opened the world of academic medicine to me. I had a wonderful research project exploring the role of the staining assays to assess urease production by *H. pylori* (a bacteria thought to cause peptic ulcers). Blaser was brilliant but not intimidating; he taught me to ask good questions and design experiments to answer them, and the laboratory was like a huge family with members from all over the world. I saw then that academic medicine was for me. I did research throughout my four years of medical school and was author on three publications by the time I graduated. I went to Johns Hopkins Hospital for my internal medicine residency. Fortunately, I had a group of fellow interns who were diverse and very bright. We became a very close group and actually wrote a paper that was published in *Clinical Infectious Diseases*.

After residency, I decided to go into private practice, primarily because I wanted to improve my independent critical thinking as an internist, and because I wanted to make money to pay off my medical school loans. I went to the University of Alabama in Birmingham and worked as a general internist. It was a wonderful experience. My patients became my family. It was fulfilling to see the "light go on" as I explained a specific disease to a particular patient. I felt as though we were winning major victories as I saw the health of my patients improve. They would bring me fruits and vegetables from their gardens or homemade gifts. Geriatric patients wanted me to meet their adult children whenever they were in town. Parents would ask me to see their older teenagers and "talk to them about . . ." I really enjoyed private practice, but over time, it lost its intellectual stimulation for me. You can only see so many patients, and my practice filled almost after the first year. By the second year, I felt as though I was performing maintenance medicine. The glycosylated hemoglobins were in an acceptable range, the blood pressures were well controlled, and everyone had their screenings. It was there that I saw sarcoidosis patients. I was amazed with the degree of devastation that this disease could cause in young people. As I read the literature, it seemed very fatalistic. I remember thinking, "Sarcoidosis has to have a cause; every disease has a cause." So I decided to return to Vanderbilt for my infectious disease fellowship and study the role of infectious agents in sarcoidosis pathogenesis. During the research portion of my fellowship, we found evidence of mycobacterial nucleic acids in our sarcoidosis specimens. It was an exciting discovery, and I used this data to successfully apply for a faculty development award from the Robert Wood Johnson Foundation. Since then, we have continued to work on the role of mycobacteria in sarcoidosis pathogenesis. I must admit that I love my job. I love designing experiments that ask important questions; I love reading articles on sarcoidosis, and writing articles that will, I hope, broaden current understanding of the disease.

As for my personal life, I met my husband, John, two weeks after completing residency. He is my best friend and strongest advocate. Four years and one dog later, we had twin boys, Cameron and Miles, who are now three. Recently, my ninety-two-year-old great-aunt moved in with us. Life is very busy at our house. I think every working mother has to prioritize the commitments on her life. For me, I realized that I have a given amount of time to influence my sons for good, so I take my commitment as a mother very seriously. To work and have a family, one must work smart as well as hard. For example, I cook dinner for the week on Sunday afternoon, because I am too exhausted after work to prepare meals. If there are work-related social meetings that request my presence, I bring my sons or stay at home. I am rarely able to work ten hours straight at my job, but I can work two or three hours while my children are sleeping in the morning, spend a few hours with them, and then go to work. (We have a nanny who stays with the kids and my aunt while I am at work.) They have my undivided attention when I get home, and we read, play baseball, and visit friends until bedtime. I work again after they have gone to bed. For me, the greatest strength is having my children on a schedule so that the day is predictable. We wrote down expectations of our nanny, and our children are flourishing. I will admit that I am working hard, but I am committing time to two things that I really enjoy, my family and my research. I appreciate that my career in academic medicine does allow me flexibility with my time.

If someone were contemplating a career in academic medicine, I would advise him or her to find a mentor and watch his or her behavior over time. I think academic medicine is wonderful—it has fulfilled my desire to teach as well as to explore; I have plenty of patient contact, and the academic environment is very stimulating intellectually. It is my dream that one day we will find the cause of sarcoidosis.

Chapter 6:
SUSAN R. WENTE, PhD
Professor of Cell and Developmental Biology

I wear multiple hats as a professor at Vanderbilt. I am chair of a basic science department in the medical school, the Department of Cell and Developmental Biology, and a member of the executive faculty. I am associate director of the Medical Scientist Training Program (MSTP), responsible for all aspects of leading this dual MD/PhD degree program. I am a course director and lecturer for multiple graduate school courses. I am a principal investigator on two National Institutes of Health (NIH) individual research project grants, and direct a laboratory of ten scientists including students, fellows, and technicians.

As chair of Cell and Developmental Biology, I spearhead our department's plan for recruiting new faculty, retaining current faculty, enriching the academic environment with seminars and retreats, expanding the PhD graduate program, and attaining state-of-the-art research resources. My roles are largely administrative and educational. I oversee the departmental business office and all aspects of the faculty recruiting. I serve as a mentor to our junior faculty, fellows, and students. I also coordinate our graduate and medical teaching efforts and personally lecture in or direct several graduate courses. I have long been committed to graduate education in the biomedical sciences and mentoring students in the MSTP career path. I have served on more than thirty-five thesis committees, have mentored the completion of dissertations in my own laboratory (five PhD and three MD/PhD), and before coming to Vanderbilt, I served as the codirector of the MSTP at

Susan Wente and her husband, Chris Hardy, toss a coin into Rome's famed Trevi Fountain.

Washington University School of Medicine. I have also recently played roles in developing new leadership and research training programs for junior faculty and postdoctoral fellows at Vanderbilt.

I completed my bachelor's degree at the University of Iowa, graduating with honors in biochemistry and garnering several significant awards. I moved to the University of California-Berkeley for graduate studies in biochemistry. My PhD dissertation with Howard Schachman, a member of the National Academy of Sciences, gave me rigorous training in protein biochemistry and biophysics. My postdoctoral fellowships were completed in New York City. First, with funding from a New York State Health Research Council Fellowship, I studied the insulin receptor with Ora Rosen at the Memorial Sloan-Kettering Cancer Center. Second, with a National Research Service Award from the NIH, I worked with Gunter Blobel at the Rockefeller University. Under Blobel's guidance, a Nobel Prize winner in medicine and physiology, I began my studies of the mechanism for highly selective, bidirectional exchange of proteins and genetic material between the nucleus and cytoplasm. I have continued research in this area of cell biology as a faculty member. I was an assistant professor in Cell Biology and Physiology at Washington University School of Medicine in St. Louis and was promoted to associate professor with tenure. I was recruited to Vanderbilt in 2002 as professor and chair of Cell and Developmental Biology. To date, I have published forty-six papers in peer-reviewed journals. Numerous awards, including the Beckman Young Investigator Award, the American Cancer Society Junior Investigator Award, and a Kirsch Investigator Award, have recognized my work. I recently served as a permanent member of the NIH study section and as editor of the journal *Molecular and Cellular Biology*. I am currently an associate editor of *Molecular Biology of the Cell*. I have organized multiple international meetings (Gordon Research Conferences, European Molecular Biology Organization [EMBO] Conferences), and have been an invited symposium speaker for numerous meetings. I am currently the principal investigator on two R01 grants from the NIH.

Although the above accomplishments may sound like I have had a clear and straight trajectory on this career path, there have been multiple points of critical decision making and change. I was born in Nebraska and raised in a town of four thousand in rural northwest Iowa. This was the biggest town in the county in the middle of "nowhere"— a three-hour drive to any major airport. I am the oldest of three, with two younger brothers. My father is one of eleven children, and the only one to graduate from college (on the GI Bill). I watched him continue his education at night school and summer schools, culminating with his doctorate in education when I was a freshman in high school. He was a business instructor at the local community college—I learned how to type by sitting in on one of his summer classes when I was in seventh grade. He retired as dean of the South Campus of the college, after also serving for many years as the chair of the Business Department. My mother was a registered nurse and worked throughout my childhood (and continues to moonlight once a month). I watched my parents balance both careers, with my mom working the three-to-eleven shift opposite my dad's teaching and school schedule. She also worked every other weekend, with my brothers and I helping with the housework and my dad in charge of meals; he is still the master of the Sunday roast! I remember always being interested in science, and I excelled at math and science in school. One of my main extracurricular activities was debate, and I loved researching the topics and designing novel (and undefeatable) proposals.

At high school graduation, my career goals could only be defined as narrow, and despite academic success (I was a valedictorian), I did not have a high degree of self-confidence. I only aimed to complete a degree that would guarantee me a good job in four years. I had ruled out nursing, as I had heard my mother talk about it too much around the dinner table and I don't think I had a personal calling for caring for the ill (most doctors in small towns are consumed with such care). I didn't have any awareness of or exposure to basic science research and had never met a professor at a university or someone who did research as a career. I don't ever recall even thinking about being an MD—and if I did, I probably dismissed it as unlikely for a small-town girl. I only applied to the local community college, Mount Mercy (a four-year nursing school, in case I changed my mind), and the University of Iowa (because of its biomedical slant on the sciences versus the agricultural

Iowa State). I received several scholarships to the University of Iowa, and with my mom's strong encouragement, I matriculated there. However, I enrolled my freshman year as a pre–dental hygiene major. I know now that I did this because it was entirely low risk, with no chance of failure. However, in the first week of classes, I reached a turning point. I was informed that pre-dental hygiene majors were not allowed to test out of freshman English, which I had, and I only had the choice of either taking the class or moving to open major (an option in which I wouldn't declare my major yet but would receive additional guidance in choosing one). I was assigned a truly progressive open major advisor who changed my enrollment to the upper level science and math courses, and eventually helped me find a home in the Biochemistry Department. Several devoted faculty, who gave me opportunities to do research and who challenged me to aim higher, surrounded me. It was there that I first learned the thrill and rigor of basic biomedical research, being given opportunities to do projects both during the school year and in the summers.

At the beginning of my senior year, I applied almost exclusively to PhD programs. However, I seriously considered accepting an offer to join the Medical Science Training Program (MSTP) at the University of Iowa. Perhaps it was a desire to leave the Midwest; however, I think my decision to pursue strictly PhD studies was based on my self-awareness that I am a perfectionist at heart, and in the end, thought I would be better off focusing on just one or the other. Ending up at Berkeley, I learned a tremendous amount and was surrounded by a world-class faculty (although there were few if any female role models) and fellow students. My graduate advisor, Howard Schachman, was very supportive of my finishing my PhD rapidly (in four years total), as well as gave me multiple opportunities to lecture, teach, and do wet-bench (laboratory) research. I chose my postdoctoral fellowship mentors with his strong guidance. Unfortunately, my first postdoctoral mentor, Ora Rosen, was terminally ill during my time in her laboratory. Although a difficult time, I realize upon reflection that I learned a tremendous amount from her. She was the mother of two grown sons and a member of the National Academy of Sciences. The circumstances also led to a second critical decision point in my career when I seriously questioned whether to start a second new postdoctoral fellowship after her death. I actually interviewed for a position at a patent law firm on Fifth Avenue in New York City, an experience that effectively motivated me to continue working at an academic research career. Gunter Blobel graciously offered me an immediate position and a highly stimulating environment from which I have not looked back. When I then moved to be an assistant professor and only the second woman hired in the Cell Biology and Physiology Department at Washington University (the first preceded me by four months), I was offered many "opportunities" to direct courses and serve on graduate steering committees. I accepted too many of these, but they all gave me credible experiences to continue moving forward in my career. However, the foundation of any career is dependent on productivity in the laboratory. A successful research program garners broad respect, and making discoveries was the element that drew me to this career in the first place.

I met my future husband in graduate school. He is a PhD scientist on the Vanderbilt faculty with his own research laboratory. I strongly feel that many, many aspects of my success are due to his support and his willingness to make sacrifices for my career. For example, when I was offered my first faculty position at Washington University, he had just started his postdoctoral fellowship at Cold Spring Harbor Laboratories. He agreed to move and take a risky non-tenure track position that eventually converted to tenure track. We have two daughters, one born in 1995, when I was an assistant professor, and the second in 1998, when I was an associate professor. Life definitely changed for us when they arrived. We made adjustments in our work schedules—no more late nights or long weekends at the bench, as this was and is strictly family time (although there have been notable exceptions to the rule, for example when grants are due and then we cover for each other). I am thought of as highly organized, and this skill remains critical. Also important was the fact that I rapidly recruited an outstanding initial group of graduate students who were highly independent. I have continued to focus on accepting students and fellows who show strong independence.

While in St. Louis, we had the great fortune of access to the university day care, open 6 A.M. to 6 P.M., Mondays through Fridays and closed for only five major holidays. My husband and I split the week so that three days a week, one of us dropped off in the mornings and the other picked up in the evenings; this reversed for the other two days and gave both of us some flexibility on at least one end of every day. In the early years, we had minimal home help for financial reasons, with only a cleaning service once every two weeks, and no family nearby. However, our family has come to the rescue on numerous occasions, flying in to help when one of us would travel to weeklong meetings or when we were overwhelmed by ear infections or fevers. My parents continue to have the girls visit them in Iowa for two weeks every summer, organized to overlap with my travel to meetings and giving them important bonding time together. In terms of work travel, I have tried very hard to stick to the "only once a month, and only nine times a year" rule. Most of my lecturing and major meetings occur in the fall, so I only accept seminar invitations at other institutions during the spring.

When we moved to Vanderbilt, my administrative responsibilities increased substantially. Thus, we decided to hire a full-time nanny who does not live with us. This has been a miracle—she helps with the homework, cooking, cleaning, and organizing after-school pickups, extracurricular activities, or coverage for our nights out. Most important, she gives us time to focus directly on the girls themselves during the few evening hours we have between dinner and bedtime, and to keep the weekends focused on activities together such as hikes and bike rides instead of chores. My husband's work responsibilities have not increased; he limits his work travel, and he is fully engaged in all aspects of our daughters' days. I have also found it very helpful to live as close as possible to work and their schools, reducing commuting time. Are there "things" I give up to have both a career and a family? Oh, yes. On the personal side I admit that I haven't been able to carve out time for exercise, that I haven't decorated all the rooms in the house perfectly, and that the scrapbooks and photo albums are two years behind. On the career side, I do not go to all the meetings in my field, and I do not have as big a lab as most of my competitors. But these are small sacrifices to make to have the time to make discoveries with my family.

To have a career in biomedical research, one truly needs to have a passion for the work. I love unraveling a problem, and especially making discoveries and knowing something first. I have also always taken great pride in mentoring others to make discoveries, from the first undergraduate I mentored at Berkeley to my current graduate students. In addition, I think one also needs a high degree of self-confidence. This is a career where the positive feedback is not constant; publications aren't always accepted, and grants aren't funded every week. Moreover, the competition is stiff and unrelenting. It is critical to be in a very supportive and innovative environment with colleagues and resources that help you be your best. One has to be innovative to balance career and family, to take advantage of opportunities, and to take risks.

As one of my colleagues at Washington University, Ursula Goodenough (the mother of five and an accomplished scientist, author, and leader of national organizations), recently wrote: "Embrace the following mantra: 'Of course, I'm going to have kids and of course I'm going to have a scientific career.'" Some would say that it was and is a risk to have a family at the same time as such a career, but my children are honestly good-luck charms. My laboratory made major discoveries in each of their birth years (one a *Nature* paper, the other a *Science* paper—my first in either journal!). I definitely think that a career in biomedical research offers some intrinsic advantages to balancing work and home life. My hours were very flexible early in my academic career when 80 percent of my effort was focused on my laboratory. In many ways, it is like running a small business; as long as the bottom line is maintained (papers are published, grants are awarded, trainees are mentored), all is well. Some weeks, this may take thirty hours, other weeks, eighty hours. However, biomedical research can be all consuming and will be if you let it. Biological problems are never all answered and because there are no set hours or appointments, it never ends. It is also important to note that it remains very difficult to stop and start a career in biomedical research or to do this career part-time. A key part of this path is learning to make conscious choices about what is important to you; the most difficult part is then accepting those choices. Although I am confronted with difficult decisions regarding my time and energy, my family provides perspective.

Chapter 7:
MARY ZUTTER, MD
Ingram Professor of Cancer Research and Pathology

My current position is Ingram Professor of Cancer Research, professor of Pathology and Cancer Biology, director of Hematopathology, director of the Residency Program in Pathology, director of the Molecular Genetic Pathology Fellowship, and codirector of the Host-Tumor Interactions Program in the Vanderbilt-Ingram Cancer Center. Each of these positions offers exciting responsibilities and presents diverse challenges.

As director of Hematopathology, I am responsible for defining the long-term goals of a division with both a heavy clinical volume and a strong research focus. We provide expertise in subspecialty areas of hematopathology, including the morphologic, flow cytometric, immunohistochemical, and molecular evaluation of pediatric and adult blood diseases. Our division consists of four faculty members, three senior and one junior. In addition, we train two hematopathology fellows each year, and train and mentor two or three residents on rotation at any given time. The clinical service is busy and includes challenging diagnostic material drawn from across the Southeast.

I also direct an active research laboratory that focuses on cell adhesion receptors, specifically the integrin receptor family. Our laboratory is focused on understanding several aspects of cell adhesion receptor function *in vivo* and *in vitro*. My laboratory is particularly interested in cancer biology with a focus on tumor-host interactions. Studies are ongoing to understand the role of integrins in interactions between tumors and the microenvironment that lead to cancer initiation, progression,

Mary Zutter backpacking on South Georgia Island off Antarctica

45

invasion, and metastasis. I am the principal investigator of three R01 research grants from the National Cancer Institute. I have served on numerous National Institutes of Health (NIH) grant review committees and program project reviews. I am currently a member of the Tumor Microenvironment Study Section of the NIH.

As director of the Pathology Department's Residency Training Program, I oversee the training of almost two dozen residents each year in anatomic and clinical pathology. I am responsible for their health and well being, as well as for their education.

As codirector of the Tumor-Host Interactions Program in the Vanderbilt-Ingram Cancer Center, I work closely with Vanderbilt-Ingram's leadership to develop a strategic plan for future directions of research. We plan retreats, organize speakers and a lecture series, and facilitate the interactions of the seven research programs within the cancer center.

I grew up an only child in Baton Rouge, Louisiana. My mother, a social worker, and my grandmother raised me. I was an extremely happy, loved, and well-protected child. My biggest worry was about growing up and losing my carefree existence. I also did not want to enter a world where every day was similar to the day before. I wanted to be entirely and completely free and responsible for my own destiny. My career choices, although somewhat roundabout, have provided me with a wonderful job that has allowed me the freedom and independence I craved as a child.

From the time I was a young child, I loved science and biology. I didn't consider medicine initially. An older cousin was a physician and a surgeon, and he was a strong influence on me for several reasons. First, he was the one in the family to whom everyone looked up. Second, he cared for my aging grandmother and the rest of my family. When I was twelve, my mother was diagnosed with colon cancer. My cousin performed the surgery, and she was cured, for which she was forever grateful. About the same time, my uncle developed Hodgkin's disease. I remember being baffled by Hodgkin's disease—I continued to ask what it was, but no one could define it for me, only that it was a malignancy and my uncle was dying from it. They couldn't even tell me which organ was involved. I became extremely curious about hematologic (blood) disorders and, although not obvious initially, I think this certainly swayed my later career decisions.

I went to Newcomb College at Tulane University in New Orleans, a liberal arts school, with the idea of "doing biology" and possibly pursuing pre-med. I was a good student and, early in my junior year, visited the pre-med advisor who had been assigned to me. He told me that I was not appropriate for medical school, that it was too hard and not a good idea for women. I never went back to see him again. Although I was uncertain and insecure about my own competence to pursue medicine, I persevered. Once in Tulane University School of Medicine, the choice of a career path was an extremely difficult one. I changed my mind many times. Direct patient contact was not as exciting to me as it was to others, so I chose pathology. Pathology as experienced as a first-year pathology resident at the University of Washington School of Medicine was also not what I had hoped, but once I found the world of blood and hematologic malignancies, I knew almost instantly what my course would be. Pamela Kidd, the director of Hematopathology, was a wonderful mentor. With her help, I started on my first clinically-oriented research projects. I fell in love with hematology and the pathology and molecular mechanisms of malignant diseases. I spent much free time and part of my hematopathology fellowship at the Fred Hutchinson Cancer Research Center. I was inspired and encouraged by E. Donnell Thomas, the director of the Fred Hutchinson Cancer Research Center at that time and subsequently a Nobel laureate.

I was fascinated with research and considered doing a postdoctoral fellowship, but one of my mentors suggested that I begin an academic career in an environment in which I would function as an attending physician on the clinical service and pursue my research interests in the laboratory of a senior investigator. I decided to take his advice. When I completed my hematopathology fellowship, I accepted my first job as instructor in the Department of Pathology at Washington University School of Medicine. I learned molecular biology during several years of working in the laboratory of Stanley Korsmeyer. During that same time, I was pursuing other projects with another mentor, Samuel Santoro. As my career developed and I matured as both a physician and scientist, I became a close colleague of Emil Unanue, chair of my department and a world-class immunologist. Although I would never have dreamed of working on immunologic questions, I now have a large NIH-funded research project on work that was initiated through conversations with Unanue. All three of these great scientists and men offered me the training and inspiration that encouraged me to continue as a clinician-scientist.

Work is not always easy, but I love my job! Every day is exciting; every day is a new challenge with new people and great ideas. I chose not to have children simply because it didn't quite fit with the rest of my life. My early love was outdoor adventure. As a kid, I spent my summers outdoors, playing in the woods and hiking. As I went through college and medical school, every vacation presented an opportunity to escape to the mountains for long backpacks and hikes, or for mountain climbing in the Sierras. The location of my residency, although I hate to admit it, was chosen primarily to put me in the mountains. My love of the outdoors and adventuring has definitely influenced the rest of my life. I spend most of my free time either adventuring or staying in shape for it. In the last ten years, I have made five exhilarating expeditions to faraway places. My first big adventure was to northern Pakistan on the Afghan-Pakistani border. In this very remote area of the world, we climbed for three weeks up the Biafo Glacier and hiked out the Hispar Glacier, the longest glacier trek outside of the Arctic and Antarctic circles. Since that time, I have climbed and trekked in Argentina, Chile, and Nepal. I was with the first group of American women to retrace Shackleton's footsteps across South Georgia Island off the coast of Antarctica. I also downhill ski, cross-country ski, hike, and climb in the Tetons and the Colorado Rockies at every opportunity. I also love to read, cook, and garden. My choice of profession has allowed me all of these amazing opportunities and offered the flexibility to pursue my passions, both professional and personal. It is a constant juggling act. I struggle with how to manage it all but enjoy the challenge.

Chapter 8:
KATHRYN M. EDWARDS, MD
Professor of Pediatrics

Currently I am a professor of Pediatrics and vice chair for Clinical Research in the Department of Pediatrics. I have several funded grants from the National Institutes of Health (NIH) and the Center for Disease Control and Prevention to conduct clinical research on vaccine development and impact.

One Christmas morning I distinctly remember finding under the tree a hand-made, starched white nurse's uniform and hat, a navy cape, and a nurse's bag with a stethoscope, tendon hammer, bandages, and thermometer. At the age of five or six, I was determined to "grow up and be a nurse." I ministered care to my favorite dolls with assorted maladies, pretending that they were real people with serious illnesses. As an only child born to parents in a small midwestern town in the late 1940s, my career choices included teaching, secretarial services, and nursing. My interest in nursing was reaffirmed by my occasional sick visits to the only physician in the community, who would diagnose my illness and then ask the nurse to deliver the curative injection of penicillin. Although I never relished the pain of the injection, my rapid return to health confirmed the importance of the therapy and of the nurse administering it.

As I progressed through school I found all of my studies exciting, but was particularly drawn to science, where the whys and hows of nature were discussed. The biologic sciences held the greatest appeal. Whether discussing the life cycles of various insects, dissecting a pickled frog, or learning the

Kathy Edwards at home with her children

functions of bodily systems, I found it all intriguing. Although my parents had not received formal college educations, they clearly articulated the importance of my education and the need for a strong work ethic. The guiding principle in my family was that whatever you did, you did it the best that you could. During my middle and high school years, I began to question whether my career options were really so restricted. If I were academically more gifted than my male classmates, why was I limited to a narrower range of career opportunities?

My academic performance in high school afforded me a scholarship to attend a midwestern liberal arts college and opened many new doors. Because my small town had little diversity, I particularly enjoyed meeting people of other ethnic and cultural backgrounds, and my classes were challenging. Chemistry was my favorite. After my first year, I considered my options and decided that my interest in chemistry could be channeled into a career as a pharmacist. I transferred to the University of Iowa College of Pharmacy, but I realized after two years that pharmacy was not for me. I wanted to be a physician and was admitted to the University of Iowa Medical School the next fall.

As one of thirteen women in my medical school class of one hundred thirty, I found the work seemingly endless. Anatomy was laborious, but microbiology and pathology were intriguing. To raise tuition money, I also volunteered for numerous research studies including a right heart catheterization; this experience provided a perspective that proved invaluable in my subsequent years as a clinical investigator. Each clinical rotation was enjoyable and I vacillated between medicine and pediatrics, but in the final assessment, I preferred the children to the "oldies." Finishing medical school in 1973, I moved to Chicago for residency and fellowship training at Children's Memorial Hospital, the pediatric service affiliated with Northwestern University Medical School. In this large inner-city teaching hospital, the experiences were varied, and the challenges abounded. During my residency years, I also pursued several research projects with the director of the Division of Pediatric Infectious Disease and decided to stay there for my infectious diseases fellowship training. Although I developed my clinical skills during the fellowship, my research training was insufficient to prepare me for an academic career.

At the end of my fellowship training, I was at a critical juncture in my career. Should I take a clinical position and dabble in research, or should I pursue additional research training and attain the tools needed for credible work? I was fortunate enough to discuss my dilemma with pediatric infectious disease giants Samuel Katz and Catherine Wilfert at Duke University, who spent several hours listening to and advising me. At the end of this discussion, my decision was clear. I needed fundamental research training, so I conducted basic research on complement proteins and bacterial opsonins in Chicago for two additional years. The work was productive and resulted in several major manuscripts. Armed with these new credentials, I accepted a position as assistant professor of Pediatrics at Vanderbilt University in 1980.

Joining the NIH-funded Vaccine Evaluation Unit at Vanderbilt, I assumed leadership of studies of bacterial vaccines, specifically *Haemophilus influenzae*, type b (Hib) vaccines. At this time, Hib was the major bacterial pathogen in young children. I established a laboratory to measure the immune responses to Hib and began a series of studies evaluating conjugate Hib vaccines in young children. Building on the initial work of my predecessor, Sarah Sell, I studied the vaccines in children at the medical center and in the community to show that the new vaccine was safe and immunogenic. The vaccine was licensed and recommended for all young children in the 1990s, and now Hib disease has virtually disappeared. I then turned my attention to studying new, less reactogenic acellular pertussis vaccines and coordinated the NIH-funded Multi-center Safety and Immunogenicity trial of thirteen new acellular pertussis vaccines produced throughout the world for infants. This was a wonderful opportunity to hone my organizational skills and to complete pivotal studies that resulted in the licensure of this vaccine as well. Serologic assays were established in my laboratory to measure immune responses to the various vaccines and were subsequently used to monitor the burden of pertussis in adolescents and adults. Since that time, I have also conducted trials for many other vaccines.

Academic medicine has provided an exciting professional life. The ability to make a difference in the lives of many children through my vaccine research has been worthwhile. To see a major pediatric pathogen such as Hib disappear is as good as it gets. Academic medicine also provides the opportunity to see interesting patients in the clinic and hospital and to put their diseases into the context of new discoveries reported in *Nature*, the *Journal of Clinical Investigation*, the *New England Journal of Medicine*, and other medical journals. To ask questions about disease processes and to translate new basic observations into understanding the pathogenesis of disease is intellectually and professionally stimulating. The drive to make a difference in an entire population of patients as opposed to a series of individuals is what stimulates me to conduct research each day.

My academic career would not have been possible without the support of my husband of thirty-five years. After a decade of serving as a corporate attorney in Chicago, my husband decided that the development of our children and others was most important to him. When our four children were young, he provided much of the primary care. He also went back to college and acquired the necessary training to be a high school teacher. Each summer, he was at home to share many adventures with the children. He has a great relationship with the children, and I marvel at all the wonderful qualities he has instilled in them.

I am very protective of my time with my family; I have attended nearly every school event and athletic competition and had a major role in assisting with homework for all the children. As a very early riser, I have gotten up several hours before my children for many years to write grants, review papers, or complete other academic tasks. When I was called back to the hospital at night or on the weekends, they understood that they needed to share their mother with others. Balancing family and work has been challenging at times, but has been worthwhile because of the personal and intellectual rewards.

Would I choose academic medicine again? Absolutely. Planning and debating clinical trials with colleagues, carefully enrolling and following subjects, reviewing and analyzing data, and reporting results that make a difference are all very fulfilling. I can't think of anything else I would rather do.

Chapter 9:
JULIA B. LEWIS, MD
Professor of Medicine

As a professor of Medicine in the Division of Nephrology at Vanderbilt, I have responsibilities in the areas of clinical research, patient care, and administration. My research interests are in the design and conduct of clinical trials in patients with chronic kidney disease. I have been principal investigator in both National Institutes of Health (NIH) and pharmaceutical industry-sponsored trials, including landmark renal trials such as the captopril trial, MDRD, AASK and IDNT. I currently have two NIH grants; in one I am the principal investigator for a multicenter trial. I have published more than one hundred articles, including a lead article in the *New England Journal of Medicine*, chaired drug safety medical boards, testified before the Food and Drug Administration, and served as primary protocol designer for international multicenter trials. I also have a large clinical practice and spend four months each year as an attending physician on the Vanderbilt medical service. I coordinate the Nephrology fellowship program and the renal clinical faculty activities.

I am the oldest daughter of two parents without college educations. My pediatrician was a woman, and although she was not specifically an inspiration to me, she certainly imprinted on me that women could be doctors. I decided to become a physician at age ten when I read the books of Tom Dooley, an American physician who provided free medical care to the poor of Southeast Asia. His books included many photographs, and in retrospect, I believe I was attracted both to

Julia Lewis and her youngest daughter at the stables

the profession and to him! When I was an adolescent, an exception was made for me to join Medical Explorers, a division of the Boy Scouts that offers the chance to explore careers in medicine. On my sixteenth birthday, this opportunity led to my first part-time job in a hospital, which baffled and to some extent upset my father, who did not want his daughter to work or to give the appearance to the neighbors that she needed to do so. For me, it was a key opportunity to gain more experience in the medical field, and I worked at the same hospital throughout high school, college, and my first three years of medical school.

I had been a straight-A student since second grade, so academically, I had many options. Despite the fact that I loved reading and learning a wide range of subjects (I had majors in biology and chemistry and minors in Russian literature and math), I was not a nerd, although my children would argue this point! I am old enough to have spent a lot of time protesting the Vietnam War and have never led a conservative personal life.

I decided during my second year of medical school to become an academic nephrologist. I found renal physiology to be the most interesting of my subjects. I knew I wanted to work in an environment where I could continue to learn, create new knowledge, advance the field, and have a varied daily experience rather than the relentless monotony of private practice. I also wanted the collegiality of an academic environment. I wanted to be a good doctor and to improve the care of patients through my research contributions. I love solving the riddle of how to balance all the factors and design the proper clinical trial.

Having seen work in a private hospital firsthand, I recognized that an academic career would offer far more flexibility for a family life than a career in private practice. I have accomplished a lot and "worked" a lot of hours. However, I feel really blessed that I have lived the adage that "if you love your work, you will never work another day." I'm also blessed that those many hours have been highly flexible. In private practice, you have to work the hours that patients come for office visits and that hospitals function; additionally, coverage is difficult, especially at the spur of the moment, and the pressure of "time being money" is always on you. By contrast, much of my work can be done after 9 P.M., when my children go to bed; I can always get coverage from my colleagues not on the hospital services; and house staff and fellows are helpful extenders. If I am on call on a weekend, I can still host my daughter's birthday party; great faculty can care for my patients in the emergency room and I can cover by phone until it is a good time for me. I can easily plan being "lunch Mom" one day a month or attend school programs. The work will generally wait.

I do not enjoy cooking, cleaning, grocery shopping, and buying postage stamps, so I do none of these activities. When I am home, I can focus solely on my children. I have a full-time nanny who has been with me for nineteen years and is now eighty-five years old. When my children were little, she cared for them during the day, and she now picks them up from school and delivers them to post-school activities. I nursed my first daughter for a year, and my nanny brought her to work to be breast-fed in my office (or on occasion in the lunch room). She also cooks and runs errands during the day but heads for home between 5 P.M. and 6 P.M. I have had a series of live-in help, as well; students who, in return for free room and board, relieve the daytime nanny as needed, do the laundry, and clean up after dinner. This help relieved the pressure for me to be at home at any precise time; however, I am home before 6 P.M. about 90 percent of the time and return to work or do paperwork at home after my children go to bed. I also take my children to school every morning.

If asked, my children would say I spend plenty of time with them, and my youngest daughter would say, "My Mom is way too involved in my school work!" I have chosen to become involved in my daughter's school in ways in which I can make a difference for the overall school and in which the time commitment fits my schedule. I sat on my daughter's school board for eight years and now sit on the capital campaign committee.

I have really never felt conflict about the balance between work and home, because it is as I designed it. I work out daily, usually in some fashion with my daughters. When they were younger, I ran with them in a carriage (my youngest daughter liked doing this until age seven or eight and sat in the carriage reading—we were a bit of a strange sight in the neighborhood). My cross trainer at home allows double duty—I have spent many a night on the cross trainer reviewing homework with my daughter and have participated in a work conference phone call from there as well. When I last chaired a drug safety medical board, we had all our phone conferences after 9 P.M.—it was easy to schedule, made us very efficient, was comfortable, and did not take me away from home.

Lastly, I strongly advocate spending plenty of time on adult life. As I remind all my friends, children will grow up; it is critically important to maintain your personal adult relationships. I am fortunate that my husband and I conduct international clinical trials together, so our joint work travel makes up a big chunk of our private time together.

I care very much about time. I run a tight schedule. I do not do tasks that are not mandatory or have no interest to me. I have no patience for inefficiency and if a meeting can happen in three minutes there is no point in it taking ten. I do not have to, nor do I, say yes to all requests. I especially decline meetings before 8 A.M., writing chapters, or participating in committees that go nowhere. By the way, meetings with Dr. Neilson fall into the "mandatory category" always and in the "of interest" category most of the time.

Chapter 10:
ELLEN WRIGHT CLAYTON, MD, JD

*Rosalind E. Franklin Professor of
Pediatrics and Law*

T he day I decided to become a doctor began in fear and helplessness.
I was six years old. A few weeks earlier, I had developed seizures and
double vision. I mostly remember having trouble learning to spell—it is
hard to do well on quizzes when the thing filled with helium in patient
rooms looks like B-B-A-A-L-L-L-O-O-O-N-N when written on the black-
board. The workup included an EEG, which was given in a small room
with a window through which I could watch my mother, and
ultimately, a pneumoencephalogram, an outdated procedure in which
my head was shaved and air was injected into my ventricles so that they
could be visualized on X-ray. No tumor was found. I remember being
very uncoordinated afterward (and my father told me that I had a
terrible headache), but I thought that the operating room was wonderful
and so decided to become a neurosurgeon. My parents encouraged me
in this dream, giving me doctor (not nurse) kits, the "Invisible Man,"
and children's anatomy and physiology books. I dissected cows' eyes in
the kitchen, much to the horror of my aunts. During adolescence, I
developed a parallel interest in ethics and took myself quite seriously.

During my freshman year in
college at Duke, I decided not to
pursue medicine for a variety of
reasons but continued my studies in
biology and ethics. I always loved
talking about what I was learning. I
remember sitting in a duck blind on
the Gulf Coast of Texas with my
father over one Christmas, explaining
to him how the *rete mirabile* allows
the ducks' feet to be cold and their
bodies warm. After graduation, I

*A young Ellen Wright Clayton
at summer camp*

started graduate school in biology at Stanford, but when I heard Paul Berg talk about the Asilomar Conference where the participants developed guidelines for the use of recombinant DNA technology, I recognized that my true interest was in science policy.

In the '70s, the path to policy was through the law, so I went to Yale Law School where I wrote my law review article on genetic counseling; it appeared more than a quarter century ago. During my last year there, I studied informed consent with Jay Katz, a psycho-analyst and law professor who opened my eyes to the complexity of the physician-patient relationship. During my clerkship with a federal judge in Alabama the following year, I had the opportunity to give grand rounds on legal issues surrounding genetic counseling to a group of obstetricians in Richmond, Virginia. Following my day talking with them about what they did and the challenges they faced, I concluded that I needed to learn more about the practice of medicine and that the best way to do that was to go to medical school. When I called the pre-med advisor at my undergraduate college to find out how to apply to medical school, they gave me the information but counseled me not to bother because I could not get in. Undeterred, I asked my mother to send my college science texts so I could prepare for the MCAT in the evenings after work. Sometimes, it is a good idea to be a bit stubborn.

Matriculating at Harvard Medical School brought a number of changes. When I entered, some of my advisors counseled me to stop my education after medical school and go straight into law education. I quickly learned, however, that taking care of patients gives you credi-bility and, for the work that I do in ethics, law, and policy, clinical work provides the discipline of reality testing. When it came time to choose a specialty, I was torn between surgery and pediatrics. I ultimately decided against the former because I concluded that, for me, being a surgeon would not leave enough time for my policy work. Besides, working with children and their families is gratifying, and just plain fun. As a result, I completed a residency in pediatrics and still practice even now.

More important, I met my soon-to-be husband, then an assistant professor of English at the University of Wisconsin, shortly before entering medical school. He had a fellowship my first year and so was able to live with me in Boston. For the next two years, we commuted on Republic Airlines and ran up horrendous phone bills, until I was able to finish medical school early and move to Madison. Our elder son was born five weeks before I began my residency. The deal was clear— I was going to be an intern, and he was going to take care of the baby while being an English professor. It seems all I did that year was work, sleep, and eat—in that order. I remember going to friends' houses during the first year; on many occasions, our son and I would sleep (one time, I fell asleep face down in my plate at the dinner table), and my husband would chat and have fun. We were incredibly fortunate to find a wonderful family day care for our son, run by a woman with a master's degree in music, who became essentially a member of our family. And we had rituals. Every night that I was on call during those three years, my husband and son would come to eat dinner with me in the hospital cafeteria. My son was actually disappointed when, at the end of my residency, he no longer got to come to the hospital in the middle of the night for food.

After residency, we came to Vanderbilt; I with entry-level jobs in the medical and law schools and my husband with tenure in the English Department. Professionally, I have had the opportunity to pursue my interest in policy in a variety of ways. I have conducted empirical research on how people actually make decisions about filing lawsuits and getting genetic testing. Most of my policy work has taken place on the national level, working with such entities as the National Human Genome Research Institute, the Institute of Medicine, the American Academy of Pediatrics, and numerous working groups, but I work in international forums and state politics as well. Airplanes, the Internet, and cell phones are my good friends. I spend quite a bit of time writing, although not nearly as much as I would like to. And I teach more than most members of the medical faculty, directing courses in the medical and law schools, running the law section in the Emphasis program, and engaging in public outreach. There have been some hitches along the way in my academic development. Five minutes before the end of class on my fourth day of teaching at Vanderbilt, for example, my three-year-old son opened a door next to the podium where I was standing, jumped in, and yelled boo. The students laughed, but I was so intent on getting through the material that I kept on teaching. I have since learned that it is important to know when to fold your cards and laugh at yourself.

Returning to the realm of the personal, our second son was born two years after we moved to Nashville. This time, we hired a nanny who stayed with him until he was three and entered Vanderbilt child care. The boys are big now. The elder is in college, and the younger is approaching high school. So how did we do it? After-school care is essential. For years, we hired college students to take the boys to sports and music practices. We drive the rest of the time. The verse in Mary Chapin Carpenter's song "He Thinks He'll Keep Her," in which the mother drives all day instantly struck a chord. My husband coached their soccer teams for years. We try to schedule around the many competitions and performances they have. I went up to my son's college recently to watch his only home track meet of the season, an event that brought home to me that there will come a time when we will not have children in the house. That still seems almost inconceivable. Meanwhile, I still bake my special chocolate chip cookies almost at the drop of the hat to lure teenagers to our house. The big screen TV, the ping-pong table, and our central location do not hurt, either.

Kids are not the only responsibility. Like many people our age, our parents are aging. I moved my mother from Houston several years ago to an apartment four blocks from our home. She is doing remarkably well, and I see her quite frequently and help with some of her chores. My husband calls his mother every day and is visiting her in Texas as I write this.

So what would I advise women who are considering a career in academic medicine? Pick your mentor, if you have one, wisely. You want someone who will tell you the truth about how you are doing and who will deal honestly with you. A wise mentor recognizes that she benefits from the success of junior colleagues. Find colleagues with whom you can *safely* share your frustrations. There are going to be some, and it can be a bad idea to share your complaints with the world. Develop good relations with the staff; life is better in every way when they are glad that you are around. And remember that academic medicine is a great job. It takes a lot to put all the personal and professional pieces together, but if it works, you get to do challenging, significant work with great colleagues, you get to help others, and you get to have close relationships with family and friends. What a gift.

Chapter 11:
ELAINE SANDERS-BUSH, PHD
Professor of Pharmacology and Psychiatry

I am a professor of Pharmacology and director of the Brain Institute at Vanderbilt University. I have been funded for many years by the National Institutes of Health (NIH) and am also the recipient of a Bristol-Myers Squibb Unrestricted Grant for Neuroscience Research. I work on serotonin biology in the brain. In 2005, I received a teaching award from the School of Medicine for mentoring students in the research setting (probably hundreds), and will shortly assume duties as president-elect of the American Society for Pharmacology and Experimental Therapeutics.

I was born and raised on a small farm in southern Kentucky. My father loved the land and raised tobacco, which was the principal moneymaking crop available to a small-time farmer. My mother worked in a factory, at first on the assembly line; after several years, she was promoted to a desk job as a receptionist. I had an older brother and a younger sister. My mother was the strong member of the family, the true head of the household.

Although neither of my parents graduated from high school, my mother was a constant proponent of education. With this encouragement and support, I was a model student, graduating second in my high school class of twenty-six. I participated in 4-H Club and Beta Club and was a cheerleader. At home, I helped milk the cows, by hand, and with general chores around the farm. My principal interests were reading, including my grandmother's nearly complete set of Zane Grey novels, which I read over and over again, and

Elaine Sanders-Bush with her daughter Kate

baseball. The Brooklyn Dodgers was my favorite team, and we listened to the games on the radio. Opportunities at my high school were limited—general biology and, amazingly, home economics fulfilled the science requirements.

I received a small grant from Western Kentucky State College and went off for my degree, only forty-five miles from the farm but light-years away in many senses, opening up a new and different world with so many fascinating opportunities for learning and so many possibilities for the future. Along the way, I discarded my dairy farmer fiancé. Several women in my extended family were schoolteachers, so this was my aspiration when I left for college. But my first true education course changed my mind—I found it quite boring and unfulfilling, so I shifted my major to biology and chemistry. My professors encouraged me to think broadly, and I decided in my junior year to attend graduate school. I considered medicine but really wanted to be a college professor, emulating the professors who were so influential in my development from a naive, reserved little farm girl to a confident adventurer.

I was undecided about an advanced degree, leaning toward either biochemistry or physiology; late in my junior year, a professor of Pharmacology at Vanderbilt University came to visit, gave a lecture, and met with rising seniors to introduce us to pharmacology and the opportunities in Nashville. I was fascinated—pharmacology was a blend of biochemistry and physiology, the two areas between which I was having a hard time choosing. So I applied to three different graduate programs—biochemistry at University of Louisville, Physiology at University of Kentucky, and Pharmacology at Vanderbilt University. I was accepted into all three programs; I chose pharmacology because of its fit with my interest and, of course, the stellar reputation of Vanderbilt University.

Now, forty-three years later, I am still at Vanderbilt. Here, I have grown up professionally and personally. I earned the PhD degree in 1967, a not-so-common accomplishment for females in those days. My female soul mate married her childhood sweetheart, had children, and dropped out of science. I was fortunate to marry a scientist, who encouraged me to pursue my career, and doubly fortunate to be offered an opportunity to join the faculty. Staying at the institution where you trained was—and still is—not the best career route. But my husband was a faculty member at Vanderbilt and not interested in moving. And Vanderbilt University was a great institution, the Pharmacology Department was top in the nation, and I never felt that I sacrificed anything by staying here. My career has flourished at Vanderbilt. I gained national and international recognition for my research on serotonin and its cellular receptors. In my current position as director of the Vanderbilt Brain Institute, I have been given the opportunity to take a leadership role, and my efforts in training, in my own laboratory, and with the neuroscience graduate program, have been the most rewarding part of my career.

We did not plan to have children; I was so focused on my research, on my career, and my husband was considerably older, so a single-parent role was likely to develop. And it was not obvious when would be a good time to have a baby. The postdoctoral and early tenure-track years are demanding and unforgiving if you don't live up to the standards. Unexpectedly, I became pregnant after eight years of marriage. We seriously considered terminating the pregnancy. My age—at thirty-five considered a danger point in those days—and my intentions both were consistent with such a decision. However, my motherly instincts kicked in and we decided that it was meant to be, and so it was the right thing to continue. My mother, who never quite understood my career decision and so wanted me to have a "normal life," was ecstatic and a great moral support.

So there I was, on June 5, 1975, a young faculty member near the prime of my career, with a beautiful, awe-inspiring baby daughter. How could I handle this? We just decided to proceed as if nothing had changed. My second graduate student was scheduled to defend his PhD thesis five days after my delivery; my obstetrician agreed that I could be induced on the scheduled delivery date if it hadn't occurred naturally, hence making it possible to schedule everything so tightly. So it was off to the hospital on June 5 after a full day in the laboratory on June 4, and Katie was born at 3:08 that afternoon. What a memorable and awe-inspiring experience! I have never, ever been sorry. So now when women ask me "When would be a good time to have a baby?" I acknowledge that there is no clear-cut, good time. Just do it, and you will find a way to make it work. For me, nurturing my baby and the entire process of parenting has been immeasurably fulfilling. I urge you—and all young women—not to forgo the amazing experience of being a mother.

My husband and I had planned to take a six-month sabbatical to perform research at the Karolinska Institute in Stockholm during the time overlapping my delivery date. We delayed the trip until after delivery, then packed our things, baby and all, and flew off to Sweden when Katie was three months old. It took courage to leave a familiar, safe environment and go thousands of miles away from my support network. But it worked out beautifully with a lot of care packages from home, mainly packed with disposable diapers. You wouldn't believe what the Swedish version was—I found many things to be far superior in Sweden, but not the diapers!

My professional development proceeded in step with my male colleagues. After a three-year postdoctoral period, I was appointed to Vanderbilt faculty as instructor in Pharmacology in 1968, then a tenure-track assistant professor in 1970. I was promoted to associate professor in 1973 and then full professor in 1980. My husband died suddenly in 1985, when Katie was nine years old. Our lives were shattered, but we recovered and moved on to a new phase. I married again two and a half years later, and we've lived happily ever after.

Soon after my second marriage, as we were considering the strategies for putting together our two careers (my second husband was also a scientist), I went on the chair-search circuit. But after I considered this and other options, we decided to stay at Vanderbilt. My research was flourishing, and my husband's career was winding down. So my husband moved from San Diego to Vanderbilt (he was here initially on sabbatical), and we actually did significant research together. Now he has retired, and we are struggling with the ravages of his advanced Parkinson's disease.

It's been a glorious forty-odd years at Vanderbilt, and I can honestly say that I was never sorry for the decisions I made that kept me here. Looking back on it now, I can reaffirm that I made the right decisions. My career has been very rewarding. I have trained many students at all stages—undergraduate, graduate, and medical students—and at the same time, my research endeavors have been recognized at the highest level, with the awarding of a ten-year MERIT grant in 1995 from the NIH. My leadership role in neuroscience training at Vanderbilt has had an impact, leaving a legacy to be remembered.

Now as I plan for retirement, my family and my heritage are major factors that will influence the journey into the next phase of my life. It is interesting to note that as I was branching out, exploring new opportunities, and building a successful career, I have stayed geographically close to home. Still today, I love the Kentucky farm, which I bought from my siblings when my parents died nearly twenty years ago. Last year, we built a wonderful getaway house on the old homestead with a garden, a pond and creek for fishing and swimming, and hiking and horseback riding trails through the woods, all of which have allowed me to get back to my farm roots.

Chapter 12:
LORRAINE B. WARE, MD
Assistant Professor of Medicine

I came to Vanderbilt in 2002 to join the Division of Allergy, Pulmonary, and Critical Care in the Department of Medicine as an assistant professor. I am in the physician-scientist track and spend 80 percent of my time doing research. My research focuses on the role of the lung lining cells in the pathogenesis and resolution of acute lung injury, and I perform both clinical and basic studies. My attending time is spent in the medical intensive care unit; this allows me to see patients who have the same problem that I study in the laboratory. Since completing my fellowship, a mentored clinical scientist award from the National Institutes of Health (NIH) has supported my research, and I am currently working to obtain independent funding.

My parents are both PhD chemists. Sometimes I wonder if they both secretly wished they had been doctors—that would explain why I was brainwashed from an early age to be one. My father, now retired, was a chemistry professor at a large university in Ontario, Canada. There were many aspects of his career that appealed to me and influenced my decision to go into academic medicine. Chief among them was the clear satisfaction and enjoyment that he derived from the challenge and creativity of scientific inquiry. I enjoyed spending time in his lab growing up. There was a comfortable camaraderie in his group of graduate students, postdoctoral fellows, and visiting scientists. I also appreciated the flexibility of his hours and the opportunity to travel that an academic career

*Lorraine Ware and
her two children*

provides. Because of my father's academic position, he was able to arrange my first research experience working in the laboratory of one of his colleagues during the summer between my junior and senior years in high school. I had a great summer working with the graduate students and postdoctoral fellows and learning a variety of laboratory methods (and washing a lot of glassware, too). That firsthand exposure to basic research cemented my interest in research as a career. From that time on, I spent every summer until medical school working in laboratories both in academic and in industry settings. My mother's career as an organic chemist was subordinated to my father's career during their marriage, contributing to their ultimate divorce. After their divorce, she did petroleum research for an oil company in southern California but never seemed to derive the same satisfaction from industrial research that my father got from his academic pursuits.

The path that has brought me to where I am now has been more circuitous than I would have predicted fifteen years ago. As an undergraduate student at Claremont McKenna College in southern California, I knew (as a result of the early brainwashing) that I wanted to go to medical school, and I knew that I was interested in research but had no firm plans beyond that. Secretly, in fact, I was very worried that I would not like being a doctor, since I had no idea what it would be like. One day while meeting with my college advisor to discuss medical school applications, I mentioned my concerns about the high cost of medical school. My advisor, knowing of my interest in research, assured me that I should not worry. He told me about MD/PhD programs and how I could have my way paid to earn two degrees. This sounded too good to be true and seemed like a natural choice for me because of my interest in research, so I applied for MD/PhD programs.

The following year I began the MD/PhD program at Johns Hopkins. The first two years were spent in medical school and included my first two clinical rotations, medicine and surgery. To my surprise and delight, I found that I really loved clinical medicine. As I turned to PhD studies for my third year, I was expecting to find the same challenge and satisfaction at the laboratory bench. Instead, I found myself back in the classroom for more, and sometimes redundant, coursework. I also ended up in a particularly dysfunctional laboratory and began to long (rather desperately) to be back on the wards. At that point I made a very difficult decision with both career and financial implications. I left the MD/PhD program (and its paid tuition and salary) and returned to medical school (with its loans) with the plan to pursue research training as a fellow rather than as a graduate student. It was my hope that by taking this route, I could let my clinical interests drive my research interests and still ultimately have a career in academic medicine. Looking back, this is one of the best decisions I have ever made. At the time, it felt like complete failure.

My training proceeded in a fairly straightforward manner for a while, and I completed medical school and started internal medicine residency at Hopkins. While I was an intern I met my husband, then a neurosurgery resident, and this introduced a whole new set of complications into my career path. Navigating one academic career is difficult—two is a monumental challenge and requires sacrifice, flexibility, and a sense of humor, with a good measure of luck thrown in. Both of our careers have been profoundly shaped by the many compromises that we have made to mesh our career aspirations with our personal lives, and we sometimes marvel that we have ended up, miraculously, right where we want to be.

Because my husband's residency did not finish until a year after mine, I took a faculty position at Hopkins for a year after residency, doing mostly outpatient internal medicine. This year was a valuable one. It taught me that there is nothing more stressful than a stack of charts in the door and a waiting room full of patients—the reason I love ICU medicine. It also gave me an extra year to choose a subspecialty and an extra year to convince my husband to leave Hopkins. As I was finishing residency and applying for pulmonary and critical care fellowships, I was longing to leave Baltimore and return to the West. My husband would have happily stayed on at Hopkins where he was firmly entrenched in a good research situation and could have made a smooth transition from resident to faculty. Instead, I cried every time he talked about staying, and we moved to San Francisco. This was no mean feat since we had to find a fellowship for me through the National Resident Matching Program (which matches applicants' fellowship requests with programs' preferences) in a city that had a fellowship opportunity for him (but no match). Somehow, things worked out, and I matched in pulmonary at University of California, San Francisco (UCSF) and my husband began a neurovascular surgery fellowship at Stanford.

Fellowship was a wonderful time for me. It was exciting to focus on in-depth training in one area of medicine and to have so much one-on-one teaching from the faculty. When it came time to choose an area for research training, I have to admit that I was guided primarily by gut feeling rather than informed decision making. I find this amusing (and a bit scary) since in retrospect, my choice of research mentors has had the greatest impact on my career of any choice I have made. Fortunately, the choice was a good one, and I ended up with the one element that I had been lacking in my quest for a career in academic medicine: a good mentor. In addition to teaching me the fundamentals of basic and translational research, my mentor has been (and continues to be) committed to helping me get my career launched; giving me the opportunity to make scientific presentations, serve as a reviewer, coauthor reviews and book chapters; and introducing me to leaders in my field. Even now that we are at different institutions, he continues to be a valued friend and collaborator.

After my first year of pulmonary fellowship, my husband's fellowship at Stanford was over, and we again faced the two-career challenge. I wanted to finish my training at UCSF, but there were no academic job opportunities for my husband in San Francisco. After some agonizing, my husband took a faculty position in Los Angeles, and for several years we maintained two households. With some creative scheduling and a lot of money spent on airplane tickets, we were able to see each other almost every weekend for the next two years. During those years of our "commuter marriage," I joked with my husband that the only way he was going to get me to move to Los Angeles was to get me pregnant— this turned out not to be a joke. Toward the end of my third year of fellowship we were expecting our first child, and it was obvious that we could not continue to live apart. However, I was really not done with my research training, had not received a fundable score on my first major grant submission, and was not yet a viable candidate for a junior faculty position. Thanks to the incredible flexibility of my research mentor at UCSF, I was able to move to Los Angeles but continue being a research fellow at UCSF—I call it my "virtual fellowship." For two years, I commuted by computer, fax, and telephone, doing bench work in my husband's laboratory and taking day trips to San Francisco once or twice a month. This allowed me to continue my training, obtain an NIH grant award, and be in the same city as my husband and baby, which was a big relief.

During my fifth year of fellowship I began to look for a faculty position. Once again, the coordination of two academic careers had a major influence on my career path. We initially looked around the country for job opportunities for the two of us but reached a dead end. Although I interviewed for a number of positions, there were no opportunities for my husband at these places. Fortunately, I was able to find an academic position at UCLA, and we finally both had jobs in the same city. Unfortunately, they were not the right jobs. If asked then, I think both of us would have said that the sacrifices we were making for our two-career family were hindering our individual careers. At that point, some much-needed luck came our way, and my husband happened to meet the chairman of Vanderbilt Neurosurgery at a meeting. This started the ball rolling for our eventual move to Vanderbilt, where we are both now happily entrenched in our research and clinical responsibilities. We consider ourselves incredibly fortunate to have found excellent academic positions at a great medical center in a city that we enjoy.

How do we make our lives work, with two kids and two academic careers? My motto has definitely become "Get the help you need." We have a full-time nanny, someone to clean our house, and our most recent acquisition, someone who comes once a month and fills our freezer with meals, saving us from the nightly ritual of coming home from work and trying to get dinner on the table while the kids vie for our attention. While it has not always been easy to give up some of the responsibility of child care and meal preparation, it makes our lives work. In addition to lots of help, it has been important to adjust expectations to be in line with reality. On the home front, as a working mother, I don't try to keep up with the stay-at-home moms. So, while

I do participate actively in my kids' education and activities, I don't volunteer to be room mother. On the work side, I find that pursuing the "triple threat" of patient care, teaching, and research sometimes leaves me feeling like I am mediocre at a lot of things instead of good at one. In reality, being a researcher enriches my teaching and clinical care and being a practicing clinician enriches my research; I just have to remind myself of this occasionally. It is also helpful to have a spouse who understands, literally, the many demands of being a physician-scientist-parent. The other motto that is useful for anyone in academics is "Just say no." Both my husband and I travel much less than we could and spend less time at work than we could. The other aspect that makes our lives manageable is the flexibility that an academic career allows. When I am not on the clinical service (most of the time), I am in charge of my schedule. I can take the kids to the doctor. If the nanny is sick, I can stay home. If an errand needs doing, I do it. This flexibility of scheduling, combined with the sheer fun of taking a scientific idea from genesis to fruition, is what keeps me firmly grounded in academic medicine.

Chapter 13:
JENNIFER A. PIETENPOL, PhD
*Ingram Professor of Cancer Research
and Biochemistry*

At the Vanderbilt University School of Medicine, I am a professor of biochemistry and lead a research laboratory of twelve investigators. We are trying to understand how tumor suppressor and cell cycle signaling pathways work in normal cells and how they are abnormal in tumor cells. Another goal of my research is to find molecular changes that occur in tumor cells and to use these alterations as targets for therapeutic interventions. This research is funded by three grants from the National Institutes of Health (NIH), two from the Department of Defense, and one from the Susan G. Komen Foundation. To date, I have published about eighty articles in a wide variety of biomedical journals, including *Science*, *Nature*, the *Proceedings of the National Academies of Sciences*, and *Cell*. In my role as a faculty member, I teach about thirty hours of medical school and graduate school lectures and on average serve on twenty-five graduate student dissertation committees a year. In addition, I have several administrative roles: associate director for Basic Science Programs of the Vanderbilt-Ingram Cancer Center, program leader for the Mechanisms of Cell Signaling Research Core in the Vanderbilt Center in Molecular Toxicology, and serve on many internal advisory boards

and steering committees for centers and programs including the Vanderbilt Physician-Scientist Program. I serve on editorial boards for five journals, and I am an associate editor for the journal *Cancer Research*. For the past ten years, I have directed and taught an annual, NIH-funded national course that introduces the latest molecular biology and translational

Jennifer Pietenpol and her husband, Mike, at home

research concepts to physicians trained in oncology-related medical specialties. I also consult for several pharmaceutical companies on anticancer drug development. Finally, I am a member of two national committees that review grant applications, one for the National Cancer Institute (Cellular and Molecular Pathobiology) and the other for the Komen Foundation (Tumor Biology).

The youngest of three children, I grew up in Rochester, Minnesota. My father was an electrical engineer at IBM, and my mother was a nurse at the Mayo Clinic. I grew up in a household where education was a top priority. My parents have strong work ethics and passed that along to me. I had my first independent position in third grade when I became a "paper boy" and had a bike route delivering the evening *Rochester Post Bulletin* to fifty customers. This was my entrée to committing to a job seven days a week and to a daily fitness routine as my route was five miles. I also had to learn financial responsibility, as I performed door-to-door collection of the fees every two weeks, kept my portion, and forwarded the balance to the publisher. After five years, I sold my paper route and began to play three seasons of athletics each academic year, an activity that I continued through my senior year in college. There is no doubt that participation in athletics was instrumental in defining the person that I am today; it gave me self-confidence and taught me a great deal about self-discipline and teamwork. Also, I received the reward of many years of training when I was named an NCAA All American athlete in my senior year of college.

My interest in science started at a young age, in part nurtured by my grandfather and father, who are outstanding, inventive engineers. Whenever I was in their company, there was always a project under way that required experimentation (usually aviation-related), and through them I became fascinated with how things worked. Whenever I saw my grandfather, he would ask, "What are you building?" I always wanted to be building or experimenting so I could give him an answer. I also had excellent teachers in junior high and high school, and I was able to participate in many extracurricular science and medically related activities offered to students by the Mayo Clinic.

After high school, I pursued a degree in biology at Carleton College. I chose a degree in science because biology and biochemistry always came easy to me, and I enjoyed analytical thinking and biological concepts. By my junior year in college, I found my love of bench and translational research sparked by both a senior thesis project as well as three summers of research in various laboratories in the Pharmacology and Cell Biology Departments at the Mayo Clinic. After being accepted to graduate and medical schools, I chose graduate school because I thought it would allow me the intellectual latitude and independence upon which I had thrived. I believe my competitive experiences in both academics and athletics were major contributors to my desire to pursue a bench-based career in cancer research. As my mother always contends, cancer research was always attractive to me because it allowed me access to an ultimate competition—the quest for a cure for cancer.

In 1986, I graduated from Carleton College and entered the graduate program in Cell Biology at Vanderbilt University with the intent of pursuing my dissertation research with Harold Moses as my mentor. At the time there weren't any multidisciplinary graduate programs, so I selected the graduate program in Cell Biology for the sole reason that I would have an opportunity to perform research under his guidance. I already had had the great fortune of working with him during my summer research internships in college when he was the chair of Cell Biology at the Mayo Clinic, before he became chair of Cell Biology at Vanderbilt in 1985. After the first week of working in his laboratory, I knew he was a great leader and mentor who was very supportive of all his personnel. I thought Moses was a mentor to follow because he was highly successful as a well-funded physician-scientist, performed outstanding science, and had a track record of motivating people and advancing their careers.

Four years later, after long days at the bench and many hours spent reading every paper in my field that I could find, I completed my dissertation research and was awarded a PhD. My efforts in graduate school led to nine publications and the opportunity to present my work at national meetings and at other institutions. Of course, the four years were not without great frustration, tears, and a temptation to quit many times. Biomedical research can be described as many frustrating months with a few really awesome days that keep you going. However, I also had a terrific mentor, supportive colleagues, and a great boyfriend (Michael Higgins, now my husband), who was a medical student and intern during my time in graduate school. Without his daily encouragement and understanding, I might not have succeeded as a graduate student at the level that I did. There was rarely an evening when Mike didn't return to the lab with me and study for upcoming exams while I worked at the bench. There were numerous times he would tell me to give it another month when a three- or six-month set of experiments seemed to be going nowhere. I can remember many a night in my last year of graduate school when I would practice various presentations in the Cell Biology conference room while Mike would try to be my audience and remain alert and interested. This coincided with his internship year in internal medicine, and on several occasions during one of my practice talks, I would find that he had crawled under the conference room table and fallen asleep. I didn't take it personally because what really mattered was that he cared enough to be there.

As I was completing my graduate studies, Mike was determining the best institutions for pursuing his residency and fellowship training in anesthesiology. We had decided before I finalized any postdoctoral fellowship decisions that the Baltimore-D.C. area offered great opportunities for both of us and we would narrow our search for our next level of training to this region of the country. We were fortunate that we both ended up training at our top choice, Johns Hopkins. I was offered a postdoctoral position at Johns Hopkins Medical Institutions in the laboratory of Bert Vogelstein, who was doing groundbreaking work in the area of tumor suppressors. A very important factor in my selection of Vogelstein as a mentor was my perception that his laboratory was pursuing outstanding, competitive science.

It quickly became apparent when I met and interviewed with him and his lab personnel that he was highly supportive of his trainees and went to great lengths to promote their careers. I wouldn't trade the time in Vogelstein's laboratory for anything, as it was the perfect environment for me. During my tenure at Hopkins, I acquired many skills important not only for performing research and for analyzing data, but also for considering the bigger picture and organizing and operating a research group. The atmosphere was extremely competitive, and success only came to those who managed multiple projects and were willing to work with great diligence. This was right up my alley; I thrived and was very productive.

About halfway through my three and a half years as a postdoctoral fellow at Johns Hopkins, I realized that I wanted to pursue a career as an independent investigator and lead my own research laboratory. Because all my training occurred in laboratories comprised of a mixture of graduate (PhD) and medical students (MD and MD/PhD) as well as postdoctoral fellows who were basic scientists and clinician-scientists, I strongly believed this mixed model was the best way to achieve medically relevant translational research. Using this model, PhD-trained students and postdoctoral fellows could acquire clinically relevant information about cancer that would promote focus on projects that were more clinically grounded. Similarly, medical students and physicians training in a basic science environment would begin to view clinical situations in a more analytical manner and gain an appreciation of what it takes to perform basic research. It was my goal to establish this model in my own laboratory.

Although I had decided to pursue an independent position at this point, I admit that the prospect of starting my own laboratory was a bit daunting. However, the thought of developing my own research program and competing in a very active area of research was too enticing. I started my job search. During that search, my mind was constantly occupied with questions: What if I can't get the offer I want? What if I can't get grants? What if I can't recruit good people? What if I get scooped? However, my efforts as a postdoctoral fellow had led to nine publications, and I was fortunate to be invited to give talks at many national and international meetings and other institutions. These accomplishments gave me confidence and led to success when I applied for faculty positions at other institutions.

Mike and I were now married and jointly pursuing junior faculty positions. I was looking for a position in a highly competitive cancer center and basic science department where I could be surrounded by outstanding colleagues and be provided with significant resources. My husband, having completed a fellowship and already a faculty member at Johns Hopkins at the time, was looking for a faculty position in an anesthesiology department where he could use his clinical skills; participate in the training of residents, nurses, and students; and expand his research in peri-operative patient care. Key to our decision making was finding an institution with superb facilities where we perceived that senior investigators and department chairs would be supportive of junior faculty and their promotion.

At the beginning of our job search, my husband and I had no idea that Vanderbilt would become a number one choice; however, it did, and we moved back to Nashville. In retrospect, we couldn't have made a better choice. My chairman, Michael Waterman, was very supportive of my career. From the beginning of my tenure, Waterman made it very clear what the expectations were of me academically and what I would need to achieve to be promoted. As a result of his leadership and additional mentorship from several other key faculty (Harold Moses, Larry Marnett, and Fred Guengerich), I was given ample opportunities and support that allowed me to be highly productive.

Obviously, starting as junior faculty with a multitude of expectations and demands, my husband and I were very busy, and our careers were very consuming. During the first five years, I was able to develop a well-funded program in cancer research and build a research group. It was a lot of work! Most exciting and enjoyable for me, I was able to advance my research on the p53 tumor suppressor protein, branch out into the field of cell cycle checkpoints, and become engaged in translational research in the cancer center. It was a great learning experience and an introduction to clinical research and clinical trials. I was able to dovetail some translational projects with ongoing research in my laboratory. I found that in addition to making mechanistic discoveries at the bench about the molecular intricacies of cells, I derived extreme enjoyment from mentoring my trainees and interacting with scientists and clinicians working in diverse areas of research. In particular, I began discovering the connections between our work and others, as well as finding links between ongoing basic science research in the cancer center and the research being done clinically. I believe my earlier training in very translational environments allowed me easily to recognize research projects that could synergize.

What excites me most about my work is the ever-increasing rate and access of knowledge in science and medicine. The advances are coming so rapidly, much more rapidly than I would have anticipated when I started as a graduate student almost twenty years ago. I now believe that our research efforts are going to have a significant impact on cancer. This promise of making a major contribution is what keeps me going. I also derive pleasure from serving in a leadership role in the cancer center and creating and sharing in bigger visions that enable more competitive cancer-based research.

Another great part of my career is the people with whom I work. Since beginning as a graduate student, I have formed wonderful and lasting friendships and collaborations on which I have relied during both good and stressful times. By far, some of the most fulfilling times in this career are the days I have the pleasure of watching my graduate students defend their dissertation research and begin their own independent careers. Already, six students have obtained their PhDs under my mentorship, and I have six more who will follow in the next couple of years. I would never have anticipated when I was a student that I would enjoy mentoring young scientists and clinician-scientists as much as I do; it is definitely one of the most rewarding aspects of my position.

During this time, my husband began to lead one of the divisions in his department, completed an MPH degree, and developed peri-operative informatics systems for the hospital. In addition to work, we enjoyed a very active and busy life outside the medical center, filled with athletic endeavors, sailing, and travel. By our fourth year in Nashville, we began another chapter of our lives with the purchase of a farm near Gallatin and began managing a menagerie of animals (horses, donkeys, cattle, horses, goats, geese and ducks). Our love of land and animals likely stemmed from our upbringing in the Midwest and inspired this experiment and adventure that we enjoyed together. My husband and I have always been fortunate in that we both accept and support each other's work habits; we share equally in chores of home-life and share fundamental personal and career goals and visions. We have always been very project-oriented and never afraid to take on a new endeavor.

After several years of taking on more and more leadership roles at Vanderbilt, it became more difficult to manage the farm and two very active faculty positions that involved an increasing amount of administrative duties. In addition, we knew we wanted to start a family and would need to carve out time for that, as well. Thus, we left the animals and land (for the time being anyway) and moved back into Nashville. With a bit more time, we were able to more easily advance our careers at the institution, with me taking on the role of associate director of Basic Sciences in the Vanderbilt-Ingram Cancer Center and my husband becoming the vice chair of his department and then interim chair.

Integral to anyone's success is a support system. For me, that system included very supportive parents, fantastic friends and colleagues, wonderful mentors, and a terrific husband. My husband and I are lucky in that our positions are very complementary in terms of time commitment, types of projects, and similar challenges. Each of us has done whatever we could to help the other during stressful times such as writing grants or preparing major presentations. We also carpool to Vanderbilt and walk or run together daily, which gives us time to talk and share ideas. It helps that we both have professions in medicine because this better allows us to share in the other's accomplishments with full appreciation of the effort required to achieve them.

My advice to young women contemplating a career path in medical investigation is "Go for it!" If you love science and medicine and the interplay between the two, the pursuit of a career in medical research will keep you engaged for life. If you don't mind competition and working hard, if you enjoy having to learn something new every day, if you are flexible enough to try new ideas, and if you have the confidence to persevere when you hit the bumps, this is the life for you. Don't worry that you haven't worked out every step in your life's plan—just do it! I never set out with the intention of holding all the positions I have or being engaged in the diverse set of activities that I am. I just knew that I loved science, so I chose to start there and work hard, and other opportunities presented themselves. As I write this, I look forward to my next opportunity as my husband and I eagerly await the birth of our son. We are ready to begin a new chapter of our lives with the challenges and joys of combining it all.

Chapter 14:
JUDY L. ASCHNER, MD
Professor of Pediatrics

I am a second-generation American and part of the first generation in my family to go to college. My current position as professor of Pediatrics and director of the Division of Neonatology at Vanderbilt University Medical Center would have surprised my greatest fan (my dad) and on many days is a marvel even to me. My career path has been somewhat circuitous, and I owe whatever success I have achieved to a supportive spouse, an innate optimism mixed with stubbornness, and a great deal of good luck.

I am first and foremost a neonatologist. Neonatology is a fast-paced, demanding critical care subspecialty. The hours are long, the patients high risk, and the Neonatal Intensive Care Unit (NICU) venue unpredictable. As a pediatric resident, you either love it or hate it. I loved it then, and I love it still. While I spend far less time these days on clinical service than I once did, I enjoy my clinical service months and still consider patient care to be what I do best. In addition to patient care responsibilities, I have a laboratory funded by the National Institutes of Health (NIH) where our investigations focus on the perinatal transitional pulmonary circulation and the vexing clinical problem of neonatal pulmonary hypertension. As I will relate later in this story, I was a late bloomer when it came to bench research and did not direct my attention to a basic science research career until nearly seven years after completion of my fellowship and four years after the birth of our fourth child. In

Judy Aschner and her family in Nashville

recent years, I have been actively involved in clinical trials in neonatology. To me, these three activities are inextricably linked and symbiotic. I am a better clinician because I do bench research, and my greatest strength in the laboratory is my understanding and appreciation of the relevant clinical issues. My passion and motivation for doing bench research are my patients and my frustration with the fundamental gaps in knowledge about the complications of prematurity and the therapies we provide in the NICU. The ultimate goal is the translation of my own research at the bench to the bedside, and the ultimate reward is making a contribution that advances neonatal care and improves outcomes for NICU patients.

I have long been passionate about education and find few activities more rewarding than teaching. It is probably this call to teach that steered me toward academia initially. I am vice chair of an international teaching organization (IPOKRaTES) that brings accomplished clinician-scientists and clinician educators to all parts of the globe, particularly to developing countries. The time and effort invested in these educational activities pay great dividends: travel, exposure to different cultures, a better appreciation for the luxury of practicing medicine in the United States, and an audience of eager and appreciative students. I have been actively involved for many years in fellowship training in Neonatology, formerly serving as a program director and currently as the chair of the Organization of Neonatology Training Program Directors (ONTPD) of the American Academy of Pediatrics. These activities are a natural outgrowth of my desire to assure that we properly train, mentor, and prepare the next generation of academic neonatologists to lead this field. Lastly, more and more of my time is spent on activities for which I have little formal training. Neonatology at Vanderbilt is a large and lucrative division, and I am finding the newer aspects of my current job description, such as budgetary and personnel management, programmatic development, and strategic planning, both challenging and rewarding.

My hometown is Troy, an economically depressed city of fifty-five thousand in upstate New York. My father was an extremely hard-working, caring, and intelligent man. He was also profoundly deaf. My father owned a small business, but this had not been his ambition in life. He had wanted to be a math teacher, a goal that was out of reach because of his hearing loss and the Depression. He was determined that his children would have a college education, and I grew up implicitly understanding that education was a privilege and the ticket to personal growth and economic success.

I grew up with modest economic resources but with an abundance of emotional support and the confidence that comes only from knowing that you are loved unconditionally. My father always told me that I could accomplish anything in life, as long as I was willing to work hard enough. I believed him.

I was an excellent student in the mediocre public schools of Troy. I had my heart set on attending Union College, a small, private liberal arts school in Schenectady. I made the deliberate decision to borrow the money needed to pay for my college education at Union and the somewhat less deliberate decision to major in biology. I had no idea what I would do with this biology degree. I considered teaching and physical therapy. I fantasized about a career in medicine but had never met a female physician. No one (including me) had ever articulated aloud that I could become a physician. That is, no one until William Roth, professor and chair of Biology, asked me at the very end of my sophomore year why I hadn't attended the "pre-med" meeting the previous evening. That was easy to answer—"Because I am not pre-med." His next question—"Why the hell not?"—was more difficult. I made an appointment to meet with him the following day.

Had it not been for the encouragement and help of Will Roth, I seriously doubt I would have applied to medical school. He helped me redesign my junior year curriculum to meet the prerequisites for the MCATs. He arranged for me to do a semester elective shadowing physicians in a community hospital. After I had been accepted at multiple medical schools, he steered me toward the University of Rochester and was extremely instrumental in my receipt of scholarship money and other financial aid. He even called to reassure my parents, whose reaction to my announcement that I had decided to become a doctor was not exactly overwhelming enthusiasm. My dad asked how in the world I was going to finance my medical education. My mother asked, "Don't you ever want to get married and have children?" She believed a career for me in medicine and her goal of becoming a grandmother were mutually exclusive. Four children and more than thirty years later, her recollection is that it was all her idea in the first place.

When I began my first year of medical school in 1977, tuition at the University of Rochester was five thousand dollars. Largely thanks to Professor Roth, I had been awarded an 80 percent tuition scholarship. Unfortunately, tuition rose dramatically over the next four years while my scholarship remained fixed at four thousand dollars. I took loans to cover the difference in medical school tuition and living expenses, adding to the debt I had accumulated during my undergraduate education. Yet, debt never impacted my career choices. I found it ironic that I made my last medical school loan payment the month that our daughter started college!

I got married during the summer between my second and third years of medical school to the love of my life, the "Israeli boy" I had met on the beach in Eilat (a city on the Red Sea at the southern tip of Israel) when I was eighteen. This was the quintessential long-distance romance, sustained over six years primarily by aerogramme and letters that typically took three or four weeks to traverse the six thousand miles between us, a fact that is hard to imagine in this age of instant messaging. A few weeks after the wedding, I started my third year, and he matriculated at the University of Rochester to complete his bachelor's degree, having spent the years I was in college serving in the Israeli Defense Forces. He had wanted to go to medical school but as a non–U.S. citizen found that to be impossible. He considered going to Italy to complete medical school. For a while I thought I would be the first to get into an American medical school and go to Italy to do residency. We owe a debt of gratitude to Mike Sheridan, then chair of the Anatomy Department at the University of Rochester who offered Miki a position as a PhD student and set the course for the rest of his phenomenally successful career as a research scientist. I think he accepted the position so that we could be together while I finished my medical education and training, not because he had his heart set on a career as a basic scientist. It was an unselfish sacrifice. Fortuitously, biomedical research suited him, and he thrived and excelled in the academic environment.

Early in my fourth year of medical school, I decided that I wanted to enter pediatrics, but the usual process of ranking programs in the match was complicated by the fact that my husband was a graduate student and we were expecting our first child in June. I was incredibly fortunate that Bob Hoekelman, then chair of Pediatrics, had a precocious interest in part-time residency positions for women in medicine. He offered me a slot in the pediatrics residency program outside the match and told me I could do my internship year over two years. I was his guinea pig and the test case for his published article describing the experience of the Rochester pediatric residency program with part-time residency training.

I graduated from medical school on May 24, waddled off the stage with my diploma, and went into labor the next day. Thanks to Hoekelman, I was able to spend the first six months at home with our daughter and did my first half of internship from January through June. My husband was a graduate student, and for two years we lived on his stipend and my half an intern's salary of eight thousand dollars per year. I remember being jealous in the supermarket of the people with food stamps. The following year, I did the second six months of internship, alternating three months off and three months on, such that I completed three years of pediatric residency training over a four-year period.

It was not originally my intention to do a fellowship, but I couldn't deny my lack of enthusiasm for continuity clinic nor the fact that I was happiest when on my NICU rotations. Halfway through my second year of residency, I was offered a fellowship position in Neonatology. My husband was a postdoctoral fellow by this time, and we were delighted to stay on in Rochester.

Up to this point, I had not experienced any gender discrimination or even acknowledged to myself that there were special challenges for women in medicine. My wake-up call occurred during my fellowship. We had hoped to have our second child during my third year of residency, but it didn't work out that way. As the time to start fellowship approached, we stopped trying. Predictably, I realized I was pregnant about two weeks into my first year of fellowship. This was not good timing. There were two first-year fellows, and we were responsible for covering the NICU 24/7. As was the tradition in Rochester at the time, we covered the NICU twenty-four hours per day for two solid weeks, followed by two weeks of covering neonatal transports. There were times during my first year of fellowship when I came into the NICU and did not go home for five or six days. My husband did the lion's share of child care and household chores that year. I was also concerned about the disparaging comments that were still directed toward a former first-year fellow who had pregnancy complications requiring bed rest, placing a disproportionate burden on her male counterpart. I so dreaded everyone's reaction that I hid my pregnancy until twenty-five weeks, at which point concealment was no longer possible.

The only time during the year when the NICU was not covered by a first-year fellow was six weeks in the spring when the second-year fellows, who spent the entire year in the lab with no call, each did three weeks as the "pretending." I cleverly arranged for them to begin that six-week stretch when I was thirty-seven weeks, the gestational age at which my daughter had been born five years earlier. I figured I would be lucky to get that far given the insane hours I was working. At all costs, I wanted to avoid asking the other first-year fellow to do any additional clinical coverage because of my pregnancy. Typical of best-laid plans, thirty-seven weeks came and went, as did thirty-eight, thirty-nine, and forty. At forty-one weeks, I presented to Labor & Delivery not in labor but announced that one way or another I was having that baby that day because I had to be back on call in eleven days. Labor was induced; I delivered five hours later and was discharged home with our son four hours after I delivered, over the objections of my obstetrician and pediatrician. I didn't want to spend even one night of my eleven-day maternity leave in the hospital.

I was back in the NICU when my son was twelve days old. If I could go back and do one thing in my life differently, it would be this. I had a fantastic child care arrangement with a friend and neighbor who had been taking care of my daughter for four years. I knew my son would be in loving hands, but I cheated myself out of irreplaceable time with my newborn child. I was a fool and as I look back on it, I am dismayed that the faculty in Neonatology allowed me to do this. I consoled myself with the knowledge that no one would accuse me of not pulling my own weight and imagined that I had somehow cleared the path for any future fellows who should dare to be pregnant during fellowship.

The chief of Neonatology had been on sabbatical during my first year of fellowship. When he returned on July 1, I was called to his office. He matter-of-factly stated that since I had had a baby during my first year, it was obvious that I could not have possibly fulfilled all obligations to the clinical service and therefore I would have to do an additional three months as a first-year fellow. I cannot describe my outrage and anger, feelings I did not keep to myself. Eventually, the other faculty spoke up, and I was allowed to begin my second year. However, the experience left a scar that took some time to heal and taught me an important lesson about perceptions versus reality. I have had the opportunity to counsel many pregnant women at various stages of their medical training and career since then. I always advise them to take every day of maternity leave to which they are entitled and not to worry about what their colleagues will think about them and their pregnancy. They will think it regardless.

I spent my second year of fellowship in the laboratory, my first true bench research experience. At this time, fellowship was only two years, but a third year was strongly encouraged in Rochester and I seriously considered staying the extra year to gain more laboratory experience. However, both my husband and I were offered tenure track faculty positions at Albany Medical College, near my hometown, and we announced we would leave at the end of my second year. The division chief refused to let me leave. Eventually a compromise was struck. Albany Medical College agreed to hold our positions until January, and I agreed to stay for an additional six months in the laboratory. I was also pregnant with our third child, due in January, so this seemed like it would work out well.

In August 1987, I ruptured membranes at twenty-one weeks' gestation. Although I knew I would lose the baby (most women with ruptured membranes deliver within a week), I could not bring myself to terminate the pregnancy. After five days in the hospital, I convinced my obstetrician to let me go home. After all, the outcome would be the same whether or not I delivered in the hospital. I spent the next ten weeks at home on complete bed rest. We had two small children at home. My husband came home in the middle of each day to be sure I had eaten and was not in labor. We were terrified, particularly when I hit twenty-five weeks, the borderline of viability in 1987. By twenty-eight weeks, I was convinced I would deliver a child with poorly formed lungs. On Halloween, at thirty-one weeks, I went into labor and delivered a critically ill, septic, preterm infant with respiratory distress, low blood pressure, and few white blood cells. No one was sure he would survive that first week. He was in the NICU until December when I took him home on oxygen and a monitor. A few weeks later, we moved to Albany. To this day, my husband avoids the NICU; it brings back too many difficult memories. I practice neonatology and communicate with parents differently because of this experience.

I spent six and a half years on the faculty at Albany Medical College, mostly practicing clinical neonatology and raising four children. Our fourth child was born during my third year on the faculty—and yes, I took my full six weeks of maternity leave! I was director of the small fellowship program there, and I dabbled at bench research whenever the busy, understaffed NICU allowed me a block of research time. By this time, my husband was a successful NIH-funded investigator. I was his editor for all manuscripts that came out of his prolific laboratory and, over time, found I was able to make some contributions beyond correcting his by-then-nearly perfect English. I was also consumed by a clinically relevant, basic research question. Infants with PPHN (Persistent Pulmonary Hypertension of the Newborn) were treated in the 1980s and '90s by hyperventilation and alkaline infusions because alkalosis selectively dilates the pulmonary vascular bed. I wanted to understand why and how the pulmonary vascular response to pH differed from the systemic vasculature. Because I could find no satisfactory answer in the literature, I logically concluded that I needed to research this question myself. I began to methodically learn the needed techniques and to write foundation grants for funding.

I also knew I would need to leave Albany, but who would give protected time to a clinical neonatologist six years out of fellowship, with less than a dozen manuscripts and no funding? Then, as fellowship director, I got a call from Wake Forest University inquiring about promising fellows interested in bench research. Half jokingly, I asked if they might be interested in a somewhat older, slightly used neonatologist. I sent my husband's and my CVs, and the rest is history. I am to this day very grateful to Wake Forest for gambling on an unproven, mid-level neonatologist armed for the laboratory mostly with enthusiasm and an idea.

Despite my enthusiasm, reasonable protected time, and an American Lung Association (ALA) grant that was funded just as I left Albany, it took longer than I thought to get things rolling in the laboratory. My very successful husband was both an inspiration and a deterrent. I saw what it took to stay funded and be productive and worried about competing for grants with better-trained PhDs who devoted full time to their research. I was fortunate to receive a grant from the March of Dimes as my two-year ALA grant ended, and shortly thereafter I wrote my first NIH grant. Times were good at the NIH; to my surprise and delight, my grant was funded on its first submission. This allowed me to reduce my clinical time from four months to three. I continued to direct the fellowship-training program and took the job of mentorship seriously.

In the meantime, my family was growing and thriving in Winston-Salem. In the early years at Wake Forest, I often worried about missing soccer games and not being the PTA volunteer of the year. However, taking stock of my children reassured me. All four were well-adjusted, happy, bright, and independent. They seemed to be as proud of me as I was of them. I have to admit that my occasionally neglected husband sometimes complained about my long hours. Far more productive, but also more efficient than I am, he rarely brought work home with him; I always did.

After ten years at Wake Forest, I was recruited to Vanderbilt to direct the division of Neonatology. It was a job about which I had never even dreamed early in my career. Any doubts I had about accepting this job melted in the face of its allure and potential. Plus, there was a bonus no other place could match: we would have all four of our children in Nashville. Our daughter was already a medical student at Vanderbilt, having preceded us here by a year. Our oldest son decided at the last minute to matriculate at the Engineering School at Vanderbilt. The two youngest were a rising junior and freshman in high school and admittedly not happy about moving. However, they are outgoing and adaptable, and they easily integrated into their new school and rapidly made new friends. My husband's academic success by any standard had been phenomenal. As I told my new chair the first time we met, my husband was the icing on the cake of this dual recruitment. He is now the Gray E. B. Stahlman Professor of Neurosciences in the Department of Pediatrics at Vanderbilt.

In retrospect, my atypical career path worked to my advantage. By the time I turned my attention to laboratory research, my children were older (youngest was four, oldest thirteen when we moved to Winston-Salem). I had honed my clinical skills, and I was craving the intellectual stimulation I found among those who asked original questions, a characteristic often stifled in the clinical setting. It is unclear to me whether I would have been a successful clinician-scientist at an earlier stage in my career and family.

I still struggle with the right balance in life. I probably spend too much time working at night and on weekends and doing household chores I don't enjoy. I should spend more time enjoying the company of my wonderful, nearly adult children. We should take more family vacations. I definitely need to figure out how to fit regular exercise and an occasional haircut into my routine. I still am not very good at saying no.

Would I do it all over again? Absolutely! Would I encourage other women to pursue a career in academic medicine? Without hesitation—just ask my daughter. I acknowledge that I have been fortunate in many ways. I have a spouse who has been a true partner in all things, and four children, all healthy, bright, and good-natured. I enjoy and am fulfilled by all the aspects of my academic career: research, clinical practice, teaching, and, most recently, shaping the future of my division. I spend my working hours in a stimulating and intellectually robust environment. What could be better than this?

Chapter 15:
JIN CHEN, MD, PHD
Associate Professor of Medicine, Cancer Biology, and Cell and Developmental Biology

I work in the Division of Rheumatology and Immunology in the Department of Medicine. My primary role is to conduct research and teach graduate students and postdoctoral fellows. My laboratory has pioneered identification and dissection of the role of Eph receptor tyrosine kinases in blood vessel formation by tumors. I have been principal investigator on grants from National Institutes of Health (NIH), private research foundations, and pharmaceutical companies. I currently have two grants from the National Cancer Institute to conduct research involving tumor-host interaction and pathological angiogenesis. I serve on grant review panels for the NIH, Department of Defense,

and Veterans Administration (VA) Merit Review, and contribute as a reviewer for a number of scientific journals. I am also actively involved in the Interdisciplinary Graduate Program and Medical Scientist Training Program (MSTP) to recruit and mentor graduate and MD/PhD students. I have mentored five of my trainees (students and fellows) in successfully obtaining their own extramural funding.

I grew up in Shanghai, a metropolitan city in China. My parents are both engineers who put emphasis on education even though we were very poor. Poor was a relative term. I considered myself privileged because our one-room apartment had electricity and shared an indoor toilet with

Jin Chen hiking at a Keystone meeting

95

some other families—many of my classmates did not have such luxury. In my elementary and high school years, China was in turmoil with the Cultural Revolution, a movement initiated by Mao Tse-Tung to drive intellectuals into camps and send millions of high school graduates to the countryside for physical labor. As a consequence, there was minimal emphasis on formal education. Most books were burned, and the ones we were allowed to read were the *Quotations from Chairman Mao Tse-Tung* (the Red Book), the stories of heroes (usually Mao's soldiers), and eight Peking operas directed by Mao's wife, Jian Qing. Fortunately, there were still many intellectuals in Shanghai and many underground books circulating among friends. My sister would smuggle those banned books to our apartment, and we all eagerly read them in turn. Fortunately, my mother also persuaded several family friends to tutor me privately. I still remember vividly those informal lessons and lively discussions that covered a variety of topics late in the night. In such an environment, I developed a thirst for knowledge.

During my last two years of high school, Mao died, and the Cultural Revolution ended. The competitive examination to enter college was restored. That was an exciting time. The whole population was optimistic and in unity. There was a sense of urgency, so much to learn but so little time. Everybody was studying day and night to make up for lost time. There was also a sense of duty and mission, for both teachers and students, to train and be trained to rebuild the country that had been badly wracked by the Cultural Revolution. In that era, girls and boys were equally encouraged to study hard and to make a contribution to our country. As a result, I was able to pass a very competitive national exam to enter a prestigious medical college in Shanghai.

My interests in science and academics were not fully developed until college. Art was my favorite subject in high school. I was good at drawing, calligraphy, Chinese ink and brush painting, and water coloring. I originally wanted to go to a professional art school, but my mother didn't think that art was a profession and encouraged me to apply to medical school. In retrospect, she was absolutely right, and I really enjoyed my medical school years. We had a rigorous basic science curriculum as well as clinical training. Toward the end of medical school, I came to realize that I couldn't "save the world" by only practicing clinical medicine. Therefore, after graduation from medical school, I entered a master of science program to focus exclusively on laboratory investigation. Although I had always been a straight-A student, working in the laboratory quickly made me realize I needed to develop my research ability. I decided to get rigorous training in the United States. While a dozen schools accepted me to enter their PhD programs, only Harvard University offered financial support, and as I was penniless in China, it was an easy decision. Attending Harvard also turned out to be one of the best decisions I've ever made in my life because it was there that I met my husband. In fact, it was the choice of a research project at Harvard that allowed me to meet Mark; he was a fellow in a laboratory next to where I did my thesis work.

My love of science intensified progressively in graduate school and subsequent postdoctoral training. Both my PhD advisor at Harvard and my first postdoctoral mentor at the Massachusetts Institute of Technology were prominent women scientists. My second postdoctoral mentor at Vanderbilt was also a distinguished scientist, Earl Ruley. He had trained a good number of women and always shown tremendous respect for committed scientists of either gender. So I was very fortunate to be nurtured in an environment that was extremely supportive during my early scientific career. With women scientists constantly around me, together with the fact that my mother always worked when I grew up, I never doubted that I could be both a woman in research and have a satisfying family life at the same time.

I enjoy work and family and usually can find time for both. I relax with lap swimming, cooking, and reading. I also do watercolors or sketching during our family vacations, best of all with Mark's father, a mathematician who has taken up painting, too. I can manage career and home in part because of my very supportive family. My husband is also a scientist who runs a laboratory down the hall from me. He understands what I need and is always ready to help (well, almost always). You might be surprised to find that in addition to yard work, he also cleans house, does laundry, and cooks breakfast on weekends.

My mother came to live with us ten years ago to look after my son when he was young. She cooks delicious Chinese food, which relieves me of cooking some of the weekday dinners. My son, who is now twelve, is also a good helper. I rarely have to worry about his homework; he is usually on top of it. In fact, he has become my living dictionary and my teacher of popular American culture (on which I need serious help), and he always beats me in Scrabble. He has become an excellent cook who makes delicious apple pies and chocolate tortes. We chose to live within walking distance of the campus to save commuting time, so Mark and I often can walk together to our laboratories in the morning.

Several important factors contributed to my success. First was learning to choose colleagues carefully and delegate tasks. After initial mistakes, I determined to recruit intelligent, independent, hard-working, and responsible trainees and technicians to my laboratory. Now I have a team of scientists who work collaboratively and efficiently as a group, allowing me to focus on the big questions in science. Plus, they bring food to the lab! Second, I have been blessed with a supportive division chief who loves science and gives me lots of latitude. Lastly, I am careful about choosing commitments, as I usually have a tight schedule. I try not to serve on more than two committees at the same time. Even though I liked my clinical work, I am happy with a focus on research.

What I like about science and academics is that I can always be at the frontier of discovery; things don't get dull and never stay the same year after year. When I was a graduate student, I cloned a gene that encodes an enzyme to repair DNA damage in bacteria and yeast. The gene was also subsequently cloned in mammalian cells, but it was not possible to delete it in the mammalian genome at the time it was discovered. However, in a few short years, gene targeting by homologous recombination was developed, and today any gene can be knocked out in mice. So there is the excitement of new discovery and new technology arriving regularly under one's very own eye!

Another attraction of science and academics is the freedom of pursuing one's creative interests. I set the course for my laboratory and study problems that are important to me. And working in the academic setting also offers the opportunity to meet many fascinating people. In fact, I have never met any scientist who was a nerd! The actual practice of biomedical research is intensely social, more so than clinical practice. You always are interacting with other scientists, and most of them are very smart and provocative. I love the opportunity to be challenged by graduate students and postdoctoral fellows both in and outside of the laboratory.

To today's young women contemplating a career path in medical investigation, I would say: "You can do it!" I would advise you to choose a sympathetic mentor to start your research life and find a good niche for your scientific interests. Recruit good people to work with you, and treat them so well that they don't want to leave. Have a supportive family. Don't be shy to ask for help, and don't hesitate to say no. Take care of yourself, and have some fun in your busy life. Enjoy your paycheck! (Scientific careers *can* pay the bills to put the kids through college.)

Having both a scientific career and a happy family life can be very rewarding. If I could start over again, I would choose to do exactly what I have done.

Chapter 16:
MILDRED T. STAHLMAN, MD
Professor of Pediatrics

I am a professor of Pediatrics and the former director of Neonatology in the Department of Pediatrics. I helped start the first neonatal intensive care unit in the United States that used monitored respiratory therapy. I have been funded for my research in neonatal pulmonary diseases and have great interest in the mechanisms and treatment of lung injury in the newborn infant. I am a past president of the American Pediatrics Society, recipient of the Virginia Apgar Award from the American Academy of Pediatrics and the John Howland Medal from the American Pediatric Society, and was elected by my peers to the Institute of Medicine of the National Academies. In 2004, I received the Distinguished Alumni Award from the Vanderbilt School of Medicine.

I was born in Nashville in July 1922, the second of two girls. Animals always surrounded me in my childhood, primarily dogs, and in the early days these were large German shepherds who were very protective and very gentle with my sister and me. When I was five, we moved further into the country, living right at the edge of town, and the number of animals in my life increased dramatically. My father, a local newspaper publisher, was a friend of the curator of the Nashville Zoo and in addition to dogs, we children each had a pony. My father and mother bought horses and, with the exception of my sister who never really cared much for riding, we rode a great deal. I spent most

*Millie Stahlman
with a friend on her farm*

of my spare time in decent weather on my little Shetland pony, riding bareback at all times except when I showed her in horse shows. My father loved animals and had many different kinds of exotic birds. I had white rats in addition to cats. We raised chickens, geese, ducks, and pheasants. I think I was probably a lonely child but didn't know it because I identified with all of my animal friends and loved being in the country with them.

I went to a county school for the first eight years and surprisingly got a very good liberal arts education. It was there I recognized that I enjoyed memorizing poetry and songs. My sister was a class ahead of me and much more studious; she always was held up to me as an exemplary student. I went to a private girls' high school and received not only a liberal arts education but also the best they had in classical education. I had declared at the age of eleven that I wanted to be a doctor and considered myself studying for medical school by the time I was in high school. My father and mother were divorced while I was in high school, and I became more of a loner than before. But I was extremely competitive, particularly in sports, both in high school and for five years in a summer camp in Wisconsin, where I learned to jump horses and to fence.

Because my family was very involved with Vanderbilt University, and because of the oncoming threat of World War II that took my father to Washington in 1940, I went to Vanderbilt's undergraduate and medical school. Under the pressure of Pearl Harbor, most of my class went to summer school and also became seniors in absentia, so that we entered medical school after less than three years of college. I loved medical school. Very bright young people who varied from the sophisticated to country boys surrounded me, but all were clever students. When I complained to my father that I was required to study in my sorority house, a noisy environment that made it impossible to concentrate, he advised me to "Tell them to go to hell." So I did and found a quieter place to study.

There were four girls in my class, and we were all treated extraordinarily evenly both by our classmates and by the faculty. We lived at that time very near the medical school, and our home was a crossroads for all of my classmates' comings and goings at all times of day and night, always finding a snack in the fridge and a Coke placed there by my generous mother. I was primarily attracted to internal medicine, which I considered the most intellectually challenging of the specialties, but when time came to apply for internships, I found that women were not accepted into internal medicine at any of the elite eastern schools such as Hopkins, Harvard, Yale, or Rochester. I could have stayed at Vanderbilt in the Department of Medicine, but I chose to go to Case Western Reserve, where I spent fifteen months as an intern, the only woman intern at that time. Again, I was surrounded by extraordinarily intelligent and friendly house officers, both interns and residents, most of the latter having come back from the war with great amounts of practical medical knowledge. They were invaluable teachers. I decided I would probably change over to pediatrics the next year and received an internship at Boston Children's Hospital, the pediatric teaching hospital of Harvard Medical School, again as the only woman intern, a lonely business in many ways.

In those days we were not paid as interns, either at Western Reserve or at Harvard, and only received room and board in the hospital and laundry of our uniforms. At that time, there were no women medical students at Harvard. The interns and residents that I met at Boston Children's were again extraordinarily sharp, and I found it very easy to make friends. Many of us, along with my friends at Western Reserve, have kept up with each other now for fifty years. The faculty was, of course, superb, but on the whole, the atmosphere in Boston was quite different from that at Vanderbilt and at Western Reserve, and medically, most of the people there felt they were superior to just about anyone everywhere else. The following year, I returned to Vanderbilt as an assistant resident in pediatrics and found I was now senior to most of the interns who had been ahead of me in medical school, but on returning from the war had to repeat their internships. It was challenging to be their resident and not alienate myself from them, but these people have also been wonderful friends for many years.

Through a stroke of good luck, Vanderbilt had a program of visiting speakers under a grant given by the Flexner family, and the chairman of the Department of Pediatrics at the Karolinska Institute was the Flexner Lecturer during my second year of residency. Arvid Wallgren and his family spent three months at Vanderbilt, and we all got to know them quite well. When they were ready to leave, I wondered whether it would be possible for me to spend a year in Sweden with Wallgren. He and my chairman, Amos Christie, managed to get an arrangement from the Rockefeller Foundation that made it possible.

This was a watershed move in my life. I had thought all along that I would probably end up in private practice, perhaps on the Gulf Coast since there were no pediatricians there at that time. However, my year in Sweden completely sold me on academic medicine. I was put in a research laboratory, although I was given three months in clinical management of a ward. A key development of that year was my purchase of a car; friends and I traveled throughout Europe for about three months. I not only learned that science could be exciting, demanding, and rewarding, I learned that there were other forms of government, other ways to practice medicine, other ways to deliver medical care, and other values, both in Scandinavia and in Europe, that I had never been exposed to before. This experience reshaped my academic and intellectual life.

When I returned to Vanderbilt, I spent the first six months in a rheumatic fever hospital in Chicago since Christie felt that rheumatic fever was one of the big challenges of pediatrics at that time, and pediatric cardiology was in its infancy, as was sophisticated cardiac diagnostic techniques and surgery. I practiced pediatric cardiology for almost four years at Vanderbilt and found it extraordinarily unsatisfactory from an emotional and professional point of view. I was paid poorly working as an instructor in Pediatrics at Vanderbilt; I made $250 a month and lived in a $100-a-month garage apartment in the backyard of somebody's nice house in Belle Meade. In those days, cardiac surgery for infants and children and cardiac catheterization were really very primitive. I was very emotionally involved with my patients and their families and had a very hard time when I lost a baby on the operating table.

In 1952, Elliot Newman came to Vanderbilt from Hopkins as a new professor of Experimental Medicine and Clinical Physiology. I got to know Newman and his family well, and by 1954, he had offered me an opportunity to take time out from pediatrics and to join the Department of Physiology under its new professor, Rollo Park, and learn the discipline. I thought I knew a fair amount of cardiac physiology, and Newman challenged me to learn pulmonary physiology by giving me the course to teach medical students that first year. I became very interested in the problems of the newborn infants who were dying untreated with what we knew then as hyaline membrane disease. For decades, sheep had been studied as a model for lung development and pulmonary disease. With Newman's blessing, we kept four sheep in a now-filled-in courtyard in the D wing of Medical Center North, and established a laboratory next to the nursery on the third floor. We watched these "little patients" grunt, turn grey, and die, with no way to modify their course. In 1954, I received a small grant from the National Institutes of Health (NIH), and with these funds I developed homemade equipment with which I could measure pulmonary function in newborns.

By 1959, I had studied a large number of normal newborns and also was beginning to do indicator dilution curves to study their transitional circulation. Newman urged me to apply for a large renewal grant that would set up a laboratory in our nursery in order to completely transform it into an intensive care unit. Using the equipment that I had developed, along with the newly available blood-gas analyzers, in 1961, we set up our intensive care unit to begin ventilating babies who we thought would die with hyaline membrane disease. Fortunately, the first one survived, and we were encouraged to continue to work on more sophisticated methods of ventilating and monitoring babies, particularly smaller and smaller preemies who had severe respiratory distress.

I began to recruit young people from abroad and from the United States as fellows. I'm afraid that I paid my fellows poorly, too, but we all were on a very steep learning curve and I think we all wound up being extraordinarily good friends and supportive of one another. They certainly are many of my best friends today, all over the world. I can't say that I was easy on students, residents, or my fellows, because I was so determined that attention to details should be primary and that the well-being of the infant and the family took priority over all other considerations.

By 1971, the NIH had developed a program of Specialized Centers of Research (SCOR), and we applied for a newborn center of research; for the next twenty-five years, we were successful in having grant support through this mechanism. We also had training grants during that time. Many of my fellows came from Sweden, and a number from France, and I think we wound up with eighty-four fellows from twenty countries by the time I turned the reins over to Bob Cotton in 1989.

From the early days, it was apparent that one could not safely transport critically ill babies between hospitals, even within the same city and certainly not from surrounding towns, without nurses who were specifically trained in the care of high-risk newborns using equipment that matched what they would be exposed to once they got to our intensive care unit. In 1974, we pushed the governor and the state legislature very hard from Memphis, Nashville, and Knoxville and managed to get Tennessee to support a regionalized newborn program. With that, we developed very extensive teaching programs in outside hospitals, and our fellows and nurses made an enormous difference in the outcome of babies who were transported to Vanderbilt through their expertise and equipment, using special drivers on our mobile intensive care units.

By 1980, I had almost complete burnout and abruptly quit directing the clinical service. I continued to do teaching rounds once a week and remained head of the division for another ten years as principal investigator of the SCOR, but I did not have direct patient care responsibility. I didn't know really quite what to do with my time, but I went to my friend and classmate, Virgil LeQuire in the Department of Pathology, and he helped me learn to do research in pathology with Mary Gray, a consummate histologist and morphologist. For the next twenty years, Gray and I were learners together, first of the developing lung and immunohistochemistry, and then of electron microscopy of the fetus and newborn in health and disease. This was a very steep learning curve and a great challenge for me to switch gears so dramatically. Gray was not only a teacher but also my best friend. She was a combination of mother figure and sister and the most gentle and truly kind person I have ever known.

In 1989, by sheer luck I met, from the University of Cincinnati, Jeff Whitsett, who had developed perhaps the most highly productive molecular biological laboratory for developing lung in this country and probably in the world. Whitsett and I have had a highly productive relationship since then, at least from my point of view. Whitsett has never chosen to work with human material nor with electron microscopy to any extent. He has allowed me to utilize each of his exciting new probes on my large library of human fetal material and material on babies with various kinds of respiratory distress and to do electron microscopy on his knockout mice. I still work with Whitsett and his group.

Since about 1995, I have worked on a model of bronchopulmonary dysplasia, a chronic lung disease in newborns and in premature baboons whose tissue was made available to me through the Southwest Biomedical Research Institute in San Antonio, which at that time was run by a former classmate of mine, Henry McGill. The main person involved with the premature baboon problems was Jackie Coalson, who had once been at Vanderbilt and knew Gray. This relationship worked into a definitive project around 2000, and for the past five years we have worked productively with Coalson and her group on the effects of retinoic acid on septation of the premature lung.

This is where we are at the present time. The important aspects of my life and career have been, first, that I am curious. I have said that should be my epitaph, and you could take it however you wanted to, but I think curiosity is the most important factor in interesting and important research. Another factor has been clearly that I have had priceless mentors and colleagues who have been friends and surrogate family on whom I could call, I believe, for almost anything. I still have many animals, including horses and dogs. I enjoy my farm and all of the wild animals and birds and wonders of nature that it provides. I will be eighty-three in July, and I still enjoy coming to work every day. It has been a fantastic and incredibly rewarding career.

Chapter 17:
LYNN MCCORMICK MATRISIAN, PHD
*Ingram Professor of Cancer Research
and Cancer Biology*

I am one of four women department chairs in the School of Medicine at Vanderbilt. Within this group, I've now been here the longest—I came to Vanderbilt as an assistant professor in 1986 and moved up the ranks both scientifically and administratively. My research laboratory has grown to between twelve and fifteen people, and I support their activities with two long-standing grants from the National Institutes of Health (NIH), projects on several multi-investigator grants, and numerous training grants. I passed my goal of one hundred publications quite a few years ago and am now concentrating on high-impact and translational research, enhancing the careers of trainees, and collaborative transdisciplinary efforts. I think I have held every administrative and service position possible as a basic scientist, including seminar coordinator, director of graduate studies, department holiday party planner, vice chair, interim chair, animal care committee member, chair of the Appointments and Promotions Committee, and so on. The Department of Cancer Biology was the first new basic science department formed at Vanderbilt in forty years, and we celebrated our fifth birthday in April 2005. We are still a small department, with six faculty from the former Department of Cell Biology and three new recruits, but we have approximately thirty secondary faculty appointees and close to sixty students in our graduate program. We believe we fill an important niche as a basic science department dedicated to bringing scientific advances to the benefit of cancer patients as rapidly and efficiently as possible.

My dream in high school was to become an airline stewardess so I

*Lynn and Paul Matrisian at
the Great Wall in China*

could travel—maybe even to a foreign country. I took French because Canada was closer than Mexico to my central Pennsylvania hometown. In tenth grade I fell in love with biology and will always be grateful to Mr. Murray for his dedication and enthusiasm for a handful of students in a small school in a town of two thousand. My parents insisted their three children get college educations, and as the oldest, I was the first to leave home. I went to Bloomsburg State College, about an hour's drive away, and when it became clear that I was better at biology than French, I went searching for a new major. Medical technology fit the bill because it was biology-oriented and there was a clear opportunity for employment at the end of the road. My senior year was spent in training in a hospital clinical laboratory, but instead of staying at that hospital for what seemed at the time to be an incredible salary of $12,000 a year, I opted for a laboratory technician job with a genetic counselor at Hershey Medical Center for $8,000 a year. It was the right decision. Roger Ladda had a research program on aging in addition to his clinical karyotyping (chromosome) laboratory, and after the first year, I asked to be transferred to the research side of the laboratory. I became addicted to the discovery process and started to take classes towards a master's degree. Paul, my future husband, was a technician in a laboratory down the hall; our first date started a romance that continues to this day. When I realized it would take forever to get an advanced degree while working, I applied to graduate schools across the country. I was not accepted to the Cell and Developmental Biology program at the University of Arizona, but undaunted, Paul and I married, packed everything we owned in a U-Haul trailer, and moved to the desert. I thought they'd change their minds when I got there; they didn't. I enrolled in the Molecular Biology program in the School of Medicine there without financial support. We pulled weeds in our apartment complex to help pay the rent, and I found a laboratory that would pay my $4,800-per-year stipend. My advisor was delighted with my technical skills, and after graduation he asked me to manage his laboratory of ten people while he went on sabbatical in Strasbourg, France. I said sure, and when he returned, I asked if there was a spot in Pierre Chambon's laboratory in France for a postdoctoral fellow. There was, but only if I could bring my own fellowship. Paul and I moved to France before the NIH budget was passed that year, resulting in all sorts of financial and administrative havoc, but in the end we stayed two and a half years and I was ready to look for a faculty position. I had an offer from a state school with secure salary support, but Vanderbilt had the best environment. I've been here ever since.

I wake up amazed that I am doing what I am doing. I had no grand plan for my life and have simply seized opportunities, taking paths that felt right despite obvious risks. I could not do it without a supportive spouse. We decided not to have children, and household tasks have generally been shared until recently. Paul took a leave of absence when I became president of the American Association of Cancer Research, a 24,000-member international professional society. He assumed all household responsibilities; took the opportunity to travel to China, Prague, and several domestic spots with me; and improved his time in sprint triathlons. He rejoices in my accomplishments and is there when I need a lift. My need to nurture is fulfilled by students and postdoctoral fellows in my laboratory. I have friends around the world, and the opportunity to see them from time to time—often in exciting new places—is an exhilarating perk of the job. I opted to pursue an administrative path without giving up scientific pursuits when I realized I felt rewarded by the accomplishments of others. The Executive Leadership in Academic Medicine (ELAM) program offered by Drexel University was instrumental in solidifying my interest in administration and in providing guidance, a network, and confidence in my ability. I try to pass some of this knowledge on to junior faculty through the Leadership Development for Junior Faculty program.

I manage my time primarily by delegating responsibility. I miss the reward of working at the laboratory bench, but it is replaced by the joy of turning a rough draft of a manuscript into a compelling story. I have senior post-docs and research faculty in my laboratory who mentor students and junior postdoctoral fellows day-to-day as they develop their own avenues of independent research. Lectures are often delegated to senior laboratory personnel with my oversight, relieving me of the time required to prepare, and providing them with valuable teaching experience. Office personnel are empowered to develop and execute programs. If things are not always done the way I would do them, it is a learning opportunity and not the end of the world. I have had to learn to say no and am comforted by the knowledge that what I am really saying is yes to something more important. I am conscious of what I need to do for myself to keep going. I don't always exercise as regularly as I need to, but I eat right, sleep well, and get regular massages. I feel better when my hair looks good, my nails are polished, and I have the right outfit to wear. When I feel good, I perform better. I know the signs when I am too close to the edge and know that I need to take a break or risk permanent damage to valuable relationships.

I'm convinced I have the best job in the world. What I am doing has the potential of making a difference to the 1.4 million individuals who will get cancer in the United States this year. I work with bright and energetic people, and I see them reach and evolve to a level they hardly thought possible. I have been to all but two states and to many countries around the world—and didn't have to serve a single bag of peanuts to do so. And they actually pay me to have this much fun!

Chapter 18:
J. ANN RICHMOND, PHD

Professor of Cancer Biology, Cell Biology, and Medicine

My major occupation at Vanderbilt is working on cancer research. I lead a team of junior faculty, postdoctoral fellows, graduate students, and research assistants into explorations into the role of chemotactic cytokines and their receptors in tumor progression. We also explore the mechanisms that are involved in the regulation of expression of these cytokines and receptors with the hope of developing new therapies for cancer patients. The National Cancer Institute (NCI) and the Veterans Administration (VA) have funded my research program for twenty-three years. I currently have two grants from the National Institutes of Health (NIH), a VA Merit Award grant, an HBCU VA grant, and a VA Senior Career Scientist Award. I serve as co-investigator on two clinical trials, one with the Food and Drug Administration and another with the NCI. In addition, I am co-investigator on a large multi-investigator grant that is developing mathematical models for cancer metastasis. I serve as vice chair for the Department of Cancer Biology, associate director

for Education for the Vanderbilt-Ingram Cancer Center, and assistant dean under Senior Associate Dean Roger Chalkley for the Biomedical Research, Education and Training Program. I feel very fortunate to have had the opportunity to work with outstanding leaders, faculty, students, and staff at Vanderbilt University School of Medicine for more than sixteen years.

I grew up in a small farming town in Arkansas as one of five daughters. My father was a lawyer/writer who attended Harvard, and my mother did not graduate from high

Ann Richmond and her daughters

113

school. Hard work and the importance of getting a good education were major family values emphasized by both my parents. This was easy because I loved learning and took full advantage of the unusually excellent school system available to me. However, it never occurred to me to aspire to become a scientist. There were no female role models in my community for this, so I majored in education with a secondary emphasis in chemistry at Northeast Louisiana University and became a high school teacher. The aspects of teaching chemistry that were most appealing to me were research papers and the science fair. During this time of working intensely with the students to develop their projects, I got a taste of teaching research and caught the research disease. I obtained a master's degree at Louisiana State University (LSU) during the summers while I was teaching. In graduate school at LSU, I became intrigued with cell biology, then after additional years of teaching, I went on to Emory University to obtain a PhD in developmental biology.

By the time I entered Emory, I was married and the mother of a two-and-a-half-year-old and a six-month-old. Needless to say, it was a busy time, but the thrill of exploring the mysteries of developmental biology and the excitement of research drove me forward. William A. Elmer was my PhD mentor, and he was a wonderful teacher, mentor, and friend. After completing my PhD, I moved on to the School of Medicine at Emory University for postdoctoral work in cancer research. Dan Rudman was my postdoctoral mentor, and his brilliance and love for research provided a rich environment for new learning. His approach was to team basic scientists and oncology fellows so they could learn from each other. This proved to provide an intensely stimulating environment. During this time I became a dyed-in-the-wool research addict. I recall Rudman saying, "Ann, research is like a disease; once you get it, you die with it." Trying to learn to balance my love for the research I was doing, raising small children, and marriage was indeed a challenge. Unfortunately, for a variety of reasons, the marriage lost out, and I became a single parent.

At this early stage in my career, a nanny was not an option within my budget. Fortunately, my mother lived nearby and was often able to assist with a sick child or to allow me to attend an important scientific conference. I utilized a combination of day care followed by wonderful college students majoring in early childhood education who were willing to do after-school care and/or summer care. Sometimes a neighbor down the street allowed the children to come to her house when I had to be present for a 7 A.M. conference. I usually had a house-cleaning service to help with some of the household chores, and the children liked to help me shop. I like to cook and garden, and we often did these activities together. This was a period of much juggling, but it was indeed a very rewarding time both with my two daughters and with the science in which I was engaged. I concentrated on these two areas of life and enjoyed a measure of success in both. I was very fortunate to have a supportive family with grandparents close by and a wonderful group of friends and colleagues who were immensely supportive. The flexibility of the research schedule meant that if I needed to be away from the research office/lab for a few hours to do something with the children, I could come back later in the day (often with the children) to complete the day's work, or I could do the reading and writing at home in the evenings after the children went to bed. We worked as a team, and they were wonderfully understanding and really great kids. They were excellent students and became involved in the usual activities: Scouts, swim team, basketball, gymnastics, church activities, soccer, art, piano, and even cheerleading. However, there was a limit to the number of activities that each could engage in for any one year. We lived in a great community where walking to school was usual and the friends and neighbors were wonderful. Looking back, I see that we were very blessed, though at times the challenge was huge!

Moving the girls to Nashville when I accepted the position at Vanderbilt proved to be a significant challenge, since the family support group was in Atlanta. They were teenagers, and I learned the hard way that moving teenagers is not easy. We got through these times, and I was again fortunate to find a great after-school nanny who was a college student at Belmont. Soon, the girls were driving and felt they were too old for a nanny. However, if I needed to travel to a scientific meeting, I had several friends through work that were willing to house-sit with the girls, thus making travel quite doable. The girls were always understanding about grant deadlines, but I made certain to be available for their activities and needs as well. During these years, they would have much preferred that I be less available. However, I stayed very much aware of all that was going on in their lives, and I tried to develop an extended family to help provide a sense of community for them in this new environment of Nashville.

Once they were in college, the road was clear for enhanced work emphasis—and at last, time for a personal life! My professional mentors during these years were many, both male and female, Hal Moses, Lee Limbird, Don Rubin, Lynn Matrisian, and Roger Chalkley, to name a few. I watched these individuals carefully from a distance, though perhaps they were unaware that I was studying their leadership skills, their style, and their concern for the individual, as well as their approach to science and building the infrastructure that is needed for research to flourish. A few years later, I married a wonderful man who is a teacher, scientist, and nature lover. Though children and work still have a high priority in my life, I am learning to make time for an adult personal relationship. I no longer work all weekend, though I usually do spend some time on Saturdays and Sundays on academic endeavors. I now leave time for hiking with him, working out together at the gym, evenings out, and spending time on our relationship. I have learned that time spent together is more important than having a home-cooked meal, and I often have some help with housework and gardening. Though I still enjoy these activities, I am able to schedule around them, or let someone else do them if time does not permit.

My husband would say that I am the queen of multitasking. I often work on the computer, talk on the phone, and listen to the news at the same time. I can work on several problems at once, one with my hands, another with my mind, and still another with my emotions. But a key component to my day is starting out centered. I rise, take time for coffee, read something inspirational, and try to get centered before going to work. Once at work, I am energized and do not like to stop until the day's work is done. I appreciate having a spouse who is supportive of my work and understanding of my schedule. We often entertain the laboratory team together, and he fully enjoys the people I work with.

Much of the passion in my life has always been found in my work. I love interacting with post-docs and graduate students in their quest to explore a hypothesis, launch out into the unknown, and answer difficult questions. Learning the intracellular circuitry that guides a cell in decision-making processes is incredibly exciting! Discovering how cancer cells become deregulated in this decision-making process and how they revert back to a more embryonic status, and then trying to develop ways to reprogram these cancer cells so that they can be induced to die, is one of my dreams and has become my quest. In many ways, I identify with Don Quixote in the "Impossible Dream"—to dream the impossible dream, to fight the unbeatable foe, to bear the unbearable sorrow, and to run where the brave dare not go . . . to reach the unreachable star!

Those of us who give so much of our lives to cancer research are often perceived as chasing windmills, but real progress is being made. Key genes have been identified that, when mutated or lost, result in tumor progression. We are learning that these genetic determinants may predict responsiveness to certain types of chemotherapy or immunotherapy, and we can often predict populations that are at risk for developing specific cancers if appropriate prevention precautions are not taken. We are learning the nature of the environmental factors that increase risk in both genetically predisposed individuals and the general population. Deciphering the signal transduction pathways that result in the growth advantage of cancer cells and the metastatic properties of these cells has often preceded the development of new therapies and new understanding of etiology (the cause or origin of disease). Participating in this quest, with the hope of reducing both cancer incidence and deaths due to cancer, has given my life more meaning and become one of my life works.

I would encourage any young woman considering a career path in medical investigation to find and examine her passion. Medical investigation and/or research often becomes your life, and this can bring much joy and satisfaction. However, one must never lose hold of the need to balance this quest equally with one's personal, spiritual, and family life in order to maintain mental health as well as inner joy. It is good to fast forward to the end of your life and look back at what you had hoped to accomplish. Did you enrich the health and joy of those around you? Did you leave the world a better place? Did you train the next generation to carry forward in a manner that is superior to that of your own generation? Did you make time for family and were you a strong role model? Did you love others well? If these are goals you want to accomplish, you will need to set your priorities and organize your life in a manner that works well for you. Many paths can take you down this road, and the shortest path is not always the most rewarding.

Chapter 19:
KATHLEEN L. GOULD, PhD
Professor of Cell and Developmental Biology

As I write this story in 2005, I have been at Vanderbilt for fifteen years and am a professor in the Department of Cell and Developmental Biology. I was hired as an assistant professor in Cell Biology along with my husband, Steve Hanks, and I have focused my efforts over the years on basic research into the mechanism of cell division. I obtained support from the National Institutes of Health (NIH) and the Searle Scholar Program initially, and after being at Vanderbilt three years, I was appointed an investigator of the Howard Hughes Medical Institute (HHMI). While NIH continues to fund a portion of my research program, the HHMI supports me, most members of my laboratory, and the majority of my laboratory's endeavors. I spend little time teaching outside the laboratory. I contribute only a few hours each year to the Interdisciplinary Graduate Program and other courses offered by my department. I have had an average of one graduate student join the laboratory per year, and twelve so far have earned their PhDs. I also train postdoctoral fellows. Although I served as interim chair of our department for a year and a half, I have had few administrative responsibilities. My only committee work is as a member of the Appointments and Promotions Committee for the Medical School and the Faculty Advisory Committee for the Medical Scientist Training Program (MSTP). The majority of my time is spent on research-related activities including reviewing; I serve on the editorial board of several cell biology journals and on NIH study sections. I have published just shy of one hundred research papers.

Kathy Gould with her husband, Steve, and family enjoying St. John's National Park

119

I grew up in San Francisco and surrounding communities. I have a half brother from my father's first marriage with whom I never lived, and a younger brother with whom I did. My parents divorced when I was five, and my younger brother and I lived with our mother. However, my father was a significant presence in our lives until his death. He was a very energetic and likeable man, despite some glaring faults that we needn't go into, and he took charge of us every Sunday. On these days, we always had fun. We did a lot of beachcombing, drinking root beer floats, biking, and, during my teen years, sailing.

My father had lost his parents before he was sixteen, and he never finished high school. Perhaps because of this, he admired people who had obtained a formal education, and he expected his children not only to attend college, but also to earn professional degrees. He, of course, had no idea what this entailed. My mother was left with the nuts and bolts of raising two children. Following the divorce, it was necessary for her to take on bookkeeping jobs she detested to make ends meet. She had left her home on a farm at the age of fourteen to attend high school and then attended music and technical school. In common with my father, she hoped for good academic results from her children but extracurricular activities, which for me included ballet and piano, were even more important to her. I pursued both seriously, and during high school, these activities kept me out of school an average of thirty days per year. I stayed home to catch up or get ahead in my schoolwork as necessary. My mother also encouraged me to read a lot of good books. I was a good student in school but not an outstanding one. Math and science were easiest for me, and I took as many of these classes as possible to allow more time for other things. I did not have much time for a social life, so every essay I didn't have to write was important to me. In addition to learning an appreciation of the arts, I took from this background a strong work ethic, good organizational skills, perfectionism, ability to multitask, and self-motivation. I am still considered by most to be compulsive, and I admit that I have a hard time doing nothing.

Following high school, I pursued a professional dance career, much to my father's dismay. An injury ended that direction soon enough, and my father enticed me to attend college by offering to pay for it. Living with my mom, I started back at a community college and earned sufficient grades to transfer to the University of California at Berkeley. With what seemed like endless time on my hands without the dance and music commitments, I sampled a wide range of courses and subjects as well as other activities typical of college students. I had fun. I became an avid backpacker. I spent lots of time on my father's sailboat. For direction in school, I had been influenced by a marine biologist at my community college to start my major in zoology. I drifted into chemistry and then eventually, in my last year, declared biochemistry as a major. I had been intending to go to medical school and volunteered at a student clinic with serious roles for volunteers. I soon began to question my pursuit of medicine because I found I had no patience with people and truly did not enjoy working with them. I preferred working with chemicals and test tubes. I took my panic attack during the MCAT as another indication that I really didn't want to go to medical school despite having the grades and scores to get in. I didn't apply.

What then? I had friends applying to graduate programs. I did not really understand what I was getting into, but I filled out applications with them (only for California schools; we were so naive it had never occurred to us to apply elsewhere) and took the Graduate Record Examination. Then my advisor insisted I get research experience and offered his laboratory as a locale. It was a horrid experience, and I quit in the midst of interviews for graduate school. What inspired me to pursue a graduate education was a series of lectures by the Nobel Prize winners Michael Brown and Joseph Goldstein near the end of the academic year. I was mesmerized by these lectures about research into the root causes of hypercholesterolemia and became convinced that biomedical research was what I wanted to do. I hoped I'd be accepted to graduate school. Just as my advisor had warned when I refused to take additional biochemistry courses instead of history courses and turned down a job requiring forty-eight-hour stretches tending separation columns in the cold room, I did turn out at college graduation to be a "jack of all trades, master of none." The only exceptional performances I turned in were in honors organic chemistry laboratory.

I learned of my acceptance into graduate school at the University of California at San Diego one day before classes began (a long story) and had a few hours to pack everything I needed into my old hatchback, find a friend in San Diego with whom to stay, and get on the road. As I pulled off the interstate and into La Jolla, my car died. I definitely began graduate school on shaky ground. I naively rotated in unsuitable laboratories and found my fit at the eleventh hour with Tony Hunter at the Salk Institute. There were few students at the Salk so I was treated more like a postdoctoral fellow from the beginning. I was given a lot of freedom to explore different directions. I threw myself into my research projects, working generally eighty hours per week, and I thrived in that open and stimulating environment. I was passionate and still am about making discoveries about the ways cells work. I made a number of important friendships during this time, got pretty good at beach volleyball, ran 10K races, was productive in the laboratory, and met Steve Hanks, who was to become my husband, in the adjoining laboratory.

I was so comfortable with my life as a graduate student that I had to be encouraged repeatedly to graduate and apply for postdoctoral fellowship positions; I did so in that order. Steve was quite instrumental in my selection of a fellowship laboratory, and I ended up moving to England to study cell cycle control with Paul Nurse, who shared in the Nobel Prize in 2001. As a testament to my tenacity, this did not rid Steve of me! There is always a concern about whether moving outside the United States for a component of one's training will make it difficult to obtain a job back in the States later. This worry occurred to many of us who joined the Nurse lab from the States. We worried about it quite a bit and, as it turned out, for nothing. No institution overlooked us because we were living abroad; it wasn't any more difficult to arrange for interviews; and each and every one of us obtained a job. We just piled up a lot more frequent flier miles.

I was very lucky with my research project in the Nurse laboratory, and after one and a half years, I was stunned to receive calls from institutions asking me to apply for jobs. At the same time, Steve was in the process of interviewing at Vanderbilt. He asked if Vanderbilt would be willing to look at me, too, given the interest from other institutions; they did, and it worked out for the both of us. I did not apply formally for any position and had given little thought to what I wanted in a job or what I would really do with my own laboratory. As with so many other steps along my career path, I was unprepared. Once again, it was dumb luck that things turned out well.

And they have. As Paul Nurse told me once, I have lived a charmed life. I have been fortunate to attract many talented students, technicians, and postdoctoral fellows to my laboratory. One of the most satisfying aspects of my job is to share in their enthusiasm for science and to participate in the development of their projects and their careers. Still, what I love best is the discovery aspect of the work; I am a data junkie and there is virtually nothing that makes me happier than good results. While I like to keep abreast of what is going on in my laboratory, I am happy to delegate responsibility and to allow people to work independently. My laboratory is well organized so that people can progress efficiently. I strive to keep frustration, theirs and mine, at a minimum.

In the other major aspect of life, family, things have gone well, too. After three years here, Steve and I began a family, and we now have two daughters, Sarah and Jessica. This was good timing career-wise since our laboratories were fairly well established at that point. The freedom to set our own schedules has allowed Steve, who is also a professor now, and me to participate fully in our children's lives. It is rare indeed that we miss a school program, play, field day, recital, or sports game. To our children, we are like any other working parents. Although they occasionally ask us to be home with them all the time, they also advise that we'd better head back to the laboratory and work on our papers or grants so that we can keep our jobs. They seem to understand perfectly, and accept, the pressures of our work.

Over the years, we've enlisted different types of help with the children. After Sarah was born, we hired a nanny to stay with her during the day. She was with us until Sarah started school. Since then, we've hired college students to pick up Sarah and Jessica from school and drive them to activities, play dates, or home. Certain babysitters have made meals, although it has not been a priority for us to find twenty-year-olds who are willing and able to cook. Since we've found college students are not 100 percent reliable, we've had opportunities to spend quality afternoons with the girls, and it is also why I try to avoid scheduling important meetings after 3 P.M. during the school year. I feel compelled to have a clean house, so to avoid spending my time cleaning we've always hired housekeepers.

In our continuing effort to balance family life with our dual careers, Steve and I both have made significant compromises and have at least tried to adopt the motto "embrace imperfection." We've lowered our expectations of ourselves and of our research programs. We reduced our traveling considerably, turning down all sorts of speaking invitations; I stopped traveling altogether for several years. Now that our children are older, we have begun to resume this. For the last few years, I have traveled between six and ten times per year and tried to limit it to one trip per month. The family stays home. I try to arrange seminar trips so that I am gone only one night. I avoid overlapping with Steve's travel commitments. Time is, of course, limiting, and Steve and I split most weekend days between us. Generally, mornings are for Steve to do what he likes, and the afternoons are for me. Much of this time is spent at the office. However, we give up some of "our time" when the other has pressing deadlines. Once or twice a week, we will also return to the office or work at home after the children have gone to bed. We try to engineer plenty of family time as well and take two substantial family vacations per year; we probably don't do as well with arranging adult-only time.

While scientific research has been the center of our adult lives because it is what we are passionate about and it pays the bills, we have many other interests. As I grow older and realize that my job is, and will be for the foreseeable future, an endless stream of papers, grants, seminars, and meetings that I will never finish, I take more time out for my other interests and my family. I continually reassess the balance between career and family and adjust as needed. I try to be picky about what I agree to do that takes me away from the laboratory or home, although there is still room for improvement. I am stressed much of the time but definitely do not feel that I have to do it all, or that anyone expects me to.

My only advice is to find a situation where you really like what you do and then do your best at it. Have confidence in your instincts, and voice your opinions. Try not to be manipulated into believing that other things are more important than what you are interested in doing. Don't overcommit your time; learn to say no to things that you do not have to or wish to do. That said, be kind to and considerate of your colleagues. You might work with them a long time. Lastly, try to be flexible and, above all, keep smiling.

Chapter 20:
GERALDINE G. MILLER, MD
Professor of Medicine and Microbiology and Immunology

I am a professor of Medicine in the Division of Infectious Diseases. Part of my time is spent providing clinical consultations on infectious disease problems to the solid organ and stem cell transplant services. This clinical focus on infections in immunosuppressed hosts has been my goal since fellowship training in immunology and transplantation. I also enjoy intermittent stints on the general infectious disease consult service so that I stay in touch with more typical infectious diseases and new and emerging infections. The remainder of my time is spent on research that combines basic science with clinical and translational. Much of this work focused on mechanisms of acute and chronic rejection of grafts, but some forays have taken me into bacterial pathogenesis, growth factor receptor signaling, and auto-immunity. My friends call this eclectic, although others think it is "poorly focused." I have found it to be fun, interesting, and a way to interact with many other scientists, clinicians, and clinical investigators. I also served on several study sections at the National Institutes of Health (NIH), which has given me an understanding of how research grants are reviewed, how the NIH works, and of various funding mechanisms. I've been fortunate to be elected or appointed to several leadership roles in the American Society of Transplantation. Currently, I serve on its Board of Directors, am a member of the Infectious Diseases Executive Committee, and will serve on its Practice Guidelines and Awards and Grants Committees. I am an associate

Gerry Miller and her husband, Tom, with friends at home

editor for the *American Journal of Transplantation*, the society's journal, which has become the premier journal in the field.

My parents immigrated to the United States from Europe in 1946, and I was born in the Bronx in 1948. World War II had destroyed my parents' families, and their lives up until then, and this had a great impact on how they viewed authority, freedom, hunger, and education. By the time I was five, my father had learned English very well, but my mother still had difficulty. My father worked and my older brother had started elementary school, so I would go to the grocery store with my mother to help read and translate the necessary transactions. This made a strong impression on me as I learned to read very early, and it reinforced my parents' emphasis on education in order to become integrated into American society. I think the early to mid-1950s must have been a safer time to raise children, because my brother and I were allowed to travel all over New York by ourselves on the subway. He was a science nerd even then, and since he was the leader we went where he wanted, usually to the science and natural history museums and the planetarium. I attribute my inability to appreciate modern art and my profound directional dyslexia to those early days in New York City— caused by my brother's dislike for art museums and the hours spent underground or in the streets between tall buildings where cues to north, south, east, and west were nonexistent.

When I was ten, we moved to Phoenix. I was excited about the move because my father promised we could have a dog and live on a farm with chickens and horses. It was a ruse. Phoenix didn't have much public transportation, and I missed the independence and freedom we had enjoyed. By the time I was in high school, I decided to go back to the East Coast for college. My high school had few students bound for college, and although my teachers tried hard to provide challenge and stimulation, it was easy to get good grades with no work. My brother graduated ahead of me and was going to the California Institute of Technology, so it seemed logical to continue our friendly sibling rivalry by going to the Massachusetts Institute of Technology (MIT).

I knew that I was poorly prepared for college, but I thought that since it had taken virtually no work to get As in high school, a moderate amount of work would suffice. I also had the idea that because MIT had decided it should admit more girls (there were fifty girls and nine hundred fifty boys in my entering class), maybe they would treat us a little more kindly. I was rapidly disabused of both notions when I failed my first physics *and* chemistry tests and was often humiliated by the graduate teaching assistants who thought it was a waste of time and effort to teach girls science. The next four years were the most challenging of my life—at times profoundly depressing and at other times pleasing when I would finally understand. My faculty advisor was Salvador Luria, who later won the Nobel Prize for his demonstration that genetic mutations in bacteria are selected and not induced. Perhaps it was his gentle encouragement, or his beautiful Italian accent, or his enthusiasm for biology that finally led me to rediscover how fascinating biology is and its innate relevance to human health and disease. At the time, and in the midst of all the brilliant scientists who were there, I felt very inadequate. I didn't think I had much to contribute to basic biology, but I had learned a great deal about science and work ethic and becoming a doctor seemed the best way to apply what I had learned.

After the grey skies and long winters of Boston, I was ready for sunshine again and decided that California would be a good place to go to medical school. I joined the second class entering the new medical school at the University of California at San Diego, along with four other women and forty-five men (a 10 percent quota for women was felt to be pretty generous then). It was an exciting experience to be part of a new school. The faculty and students were committed to each other's success. They knew all our names, there was no place to hide, and they wanted us to be outstanding. As a third-year student, I saw a Vietnamese boy with malaria. I went to the laboratory with the infectious diseases attending physician late at night and saw the parasites on the blood smear. The boy was so ill, I thought he might die. A few hours after he received his first dose of chloroquine, he was cheerfully asking to go home. My career had been decided.

The infectious diseases faculty at San Diego was committed to research, and their encouragement prompted me to go to the NIH for an infectious diseases fellowship after I finished my residency.

After finishing clinical training in infectious diseases, I did research training in the laboratory of David Sachs, a pioneer in transplantation immunology. It took some time to figure out how to integrate my research training with clinical infectious disease. I realized that training in infectious diseases and immunology was a real bridge for understanding the relationship between infection, immunity, and the consequences of immunosuppression. With the growing population of transplant patients receiving new immunosuppressants, the need for infectious diseases specialists with expertise in immunosuppression and pharmacologic interactions between these drugs and anti-infection agents has continued to increase.

The most wonderful event during my fellowship was meeting and marrying my husband, a native Tennessean. He is also in academic medicine and has been understanding and supportive of whatever I wanted and needed to do during my career, whether working late seeing patients, doing experiments, or the pressure of grant deadlines. His gentle, laid-back style is the perfect balance to my somewhat obsessive-compulsive, perfectionist personality. He reminds me of what is really important in life, cheerfully tolerates whatever turns up for supper (or doesn't), and is generally oblivious to the state of my housekeeping or lack thereof. We arrange our lives to help each other with whatever demands may be at the moment. Our move to Tennessee from Texas prompted a major lifestyle change. We moved to a little "farm" just a half hour from Vanderbilt; that was quite a change for a city girl. Since coming here, my husband has helped me learn about country life, horses, dogs, cats, deer, wild turkeys, "critters," tractors, bush-hogs, and various farm implements, some of which I can actually now use. The nice people at the Williamson County farmer's co-op call me "Miss Gerry," and I no longer find that strange. All in all, I think my career in academic medicine has given me much to be thankful for—an extraordinary person with whom to share my life, the ability to make a difference in people's lives while doing what I enjoy, and the opportunity to meet and work with great colleagues all over the world.

Chapter 21:
JULIE A. BASTARACHE, MD
Instructor of Medicine

A few short years ago, no one could have anticipated that I would be studying the intricacies of tissue factor biology as a physician-scientist, yet here I am, just finishing my third year of a pulmonary and critical care fellowship at Vanderbilt University and beginning the transition from fellow to junior faculty. During the last two years of my fellowship, I have spent 75 percent of my time performing research and 25 percent on clinical duties. The same will hold true during my next two to three years as an instructor. My journey, like that of so many others, was somewhat circuitous.

I grew up in western Massachusetts, the oldest of six girls. My father, who never went to college, sold insurance; my mother, who has a bachelor's degree in English, was trained as a social worker but chose to stay home to raise her children. For reasons that I do not completely understand, my parents were obsessed with education. They enrolled my sisters and me in special courses to learn everything from Chinese to basket weaving to computer programming. Early in my childhood, I became interested in science and would do extra science homework just for fun. The famous television scientist Mr. Wizard was my hero. I was fascinated by the knowledge that he possessed, enabling him to seemingly change the world, or at least create fire from white powder and a drop of water. Inspired by him, I would spend long afternoons in my backyard playing with my chemistry set.

Julie Bastarache and friends

I remained curious about the world throughout high school, which is when I decided that I wanted to be a research scientist. During high school, I had a summer job waiting tables at Abdow's Big Boy restaurant. Oddly enough, my job waiting tables strongly influenced my decision to become a physician. I was an introverted adolescent and was apprehensive about the amount of human contact that being a waitress would require. To my surprise, I loved waiting tables. Taking care of people was challenging but rewarding at the same time. I began to question whether I really wanted to spend the rest of my professional life sequestered in a lab with a chemistry set, as I had imagined in my backyard as a child. Perhaps Mr. Wizard had overlooked something.

Ultimately, I decided that I would rather be a doctor caring for people than a research scientist caring for rats. In retrospect, like most adolescents, I was very naive and knew nothing about being either a research scientist or a physician. Nevertheless, I think I understood at a basic level that a physician, like a waitress, could be both challenged and rewarded by helping others. Hence, I enrolled in the College of the Holy Cross in Worcester, Massachusetts, as a biology major in the pre-medical program. The first two years of college were typical for any pre-medical student and filled with biology, organic chemistry (which I secretly loved), and physics.

During my third year of college, I took the advice of my college advisor and decided to augment my application for medical school. I enrolled in a program at Ohio State University to conduct research during the summer. The summer program was my first introduction to research, and I was very apprehensive, especially after my mentor sent me some of the lab's publications to read before my arrival. My project involved studying the promoter region of the herpes virus genome, about which I had a very rudimentary understanding. The program introduced me to pipettes, gels, transfection, polymerase chain reaction, cell culture, and various other procedures and protocols, but I had no input concerning the direction of the project. I learned two things about my future career and about myself that summer. First, I found the day-to-day lab work relatively easy and straightforward. Second, I no longer had any interest in research. Armed with this new knowledge, I applied to medical school.

To my utter amazement and delight, Vanderbilt Medical School accepted my application, and I moved to Nashville in August 1995.

During my first semester, I became so immersed in anatomy, biochemistry, and pharmacology that I never gave research a second thought. During my second semester, however, I learned about a summer research project that was mandatory for my class. I was a little dismayed at the thought of spending another summer in the laboratory. Therefore, instead of studying the herpes virus genome, I chose to do something different: brain surgery on rats. Once again I joined a laboratory where I was told what to do and had no input on the project. I now know that, at that stage of my training, I should not have had any input, but at the time, my lack of input influenced my attitude towards research.

That summer, I learned surgical techniques, histology, and how to work with animals. Like my previous experience, the techniques were easy to learn, but this time I realized that the actual experimental design was more difficult. After reflecting on these research experiences, I found they had actually relieved my anxiety about working in a laboratory, which can be an intimidating place for a novice. I learned that the true challenge of research is the thought and planning behind the experiment and not the actual experiment itself. I also realized that I probably was not very excited about research because neither of these first two projects was really mine. I suspected that my attitude would probably change if I had the opportunity to design the experiments myself. Little did I know that I would soon have such an opportunity.

My thoughts about research drifted to the back of my mind as I concentrated on finishing medical school and my residency in internal medicine. During residency, I became interested in pulmonary and critical care medicine as a specialty and was very excited when Vanderbilt chose to accept me into its program. As part of the fellowship program, Vanderbilt requires all fellows to perform eighteen months of research. I viewed this as an opportunity to work on a project that I would design from the beginning. I decided that even if I went into private practice, this short research stint would be a challenging and unique opportunity to work on a problem of my own.

During my first year of fellowship, I began to consider seriously the research project in which I would participate. I had many options in both clinical and basic science fields, and choosing between these two was very difficult. Initially, I thought that I would choose clinical research because my prior laboratory experiences had been under-whelming. But as I watched the progress of the fellows ahead of me, I noticed that those who chose clinical research soon became frustrated with institutional review boards, consent forms, low patient enroll-ment, and other obstacles that can distract one from actual investigation. I questioned whether eighteen months was enough time to develop a clinical research project that would allow me to have as much input as I wanted.

The fellows conducting basic research, on the other hand, performed experiments immediately and were involved in very specific projects in which they had significant input. Plus, unlike the clinical fellows, many basic research fellows seemed to have data soon after the start of the project. Granted, only the research fellows and their mentors cared about the data, but it was data nonetheless. I then decided that if I were going to give research a serious try and have a project that was really my own, I would have to choose basic science research.

My next big (and, as I later learned, my most important) decision was to choose a mentor. As noted, I had many great options, but finally chose to work with Lorraine Ware, a relatively young faculty member who was studying coagulation abnormalities in acute lung injury. I met with my future mentor several times during my first year of fellowship and became excited about working with her. She was the prime example of a successful and happy physician-scientist who had effectively balanced research and clinical practice. Her enthusiasm was contagious. She was my new Mr. Wizard or, more properly, my first Ms. Wizard.

Before I could join her, however, I had to overcome one problem: Ms. Wizard studied coagulation proteins. Like every other medical student and resident whom I know, I despised the coagulation cascade. I almost decided to do something else, but the more experienced faculty insisted that this young physician-scientist would be an outstanding mentor and that the specific project my mentor and I pursued would be less important than the experience of working together. This advice is the best that I have received in my short career and the best that I can share with someone else. The truth is that with an inspiring mentor, a promising idea, and an inquisitive mind, you can quickly become interested in just about anything. Three years ago, I would develop palpitations and slight nausea at the mere mention of tissue factor, the initiator of the extrinsic coagulation cascade. Today, I think it is the most fascinating protein on earth.

I have been in the lab for the past two years working on my tissue factor project. During those two years, I have also completed six months of clinical time. I love the variety. I am never bored. The opportunity to go back and forth between the lab and the hospital ultimately makes me better at both clinical medicine and research. For me, the bench-to-bedside paradigm is a reality. In some ways, I am still playing with my chemistry set in my backyard in Massachusetts and waiting tables at the Big Boy, but now I know that the rewards of one complement the other. Being in the hospital reminds me of why I am working so hard on my experiments and gives me new ideas for my research. After spending a month on clinical service, with its demands on my emotions and my time, I happily return to the laboratory, where every day seems more predictable. At the bench, I have developed critical thinking skills that I can use when I am taking care of patients, making me a much better clinician. The pace of the hospital can be frenzied, but in the laboratory I set my own pace and my own schedule. As a critical care physician, I crave the intensity and the unpredictability of the intensive care unit, but I am certain that I would burn out if I treated patients full-time. Likewise, I would not want to be in the laboratory full-time; I went to medical school to care for patients. Of course, that is also the ultimate goal of my research.

Even though my research career is in its infancy, I have learned much about academic medicine and the role of a physician-scientist. Two years ago, I had no idea what academic medicine entailed, nor did I think I would enjoy it so much. Despite my brief time in research, I have gained a significant amount of knowledge concerning tissue factor biology. I actually have people asking me questions and seeking my advice about the coagulation cascade, which feels great. Never before in my life have I felt like I possessed a unique knowledge that would lead my colleagues to seek my expertise. The feeling is empowering, especially as a woman in the male-dominated field of pulmonary and critical care medicine.

Basic science research is also much more social and collegial than I had anticipated. I was afraid that I would be sequestered in some dark room all day, but actually I interact with different people every day. I also get to travel to exciting places to attend international meetings where I can present my work and meet all of the people whose papers I have been reading. Each day is a new challenge, and I have watched my project grow over time and move one step closer to my ultimate goal of adding to the body of scientific knowledge and perhaps, if I am very lucky, eventually changing the way that we care for patients.

Chapter 22:
HEIDI E. HAMM, PHD

Earl W. Sutherland Professor of Pharmacology

I came to Vanderbilt in 2000 as the chair of the Department of Pharmacology. For more than twenty years, I've focused my research efforts on understanding G proteins, a class of cell membrane proteins that transfer signals across the cell membrane. My research has uncovered many details about how G proteins work—how they turn on and off and how they interact with receptor proteins and with the effector proteins that are next in the message chain. Because G proteins are involved in so many physiological processes, my laboratory studies the mechanisms by which G protein-coupled receptors (GPCRs) activate G proteins in a variety of cellular systems that span neuro-science, cardiovascular biology, and drug discovery.

I was born into a medical family, the oldest of five children, while my parents were in school at Loma Linda, California. Some of my early memories were of peeking in the windows of my parents' home office in a small town in Kentucky. My parents were partners in a very small medical practice; Dad was a general practi-tioner and my Mom was his office nurse and midwife. I was familiar with medical things. If I was ever injured, I would be bustled into the office, and my Dad would sew me up. I never had any question about where babies came from because I saw them delivered right in the office.

Soon after they moved to Kentucky, my Mom had gone to a midwifery school. She went for six months, and during that time my aunt came to take care of me. After my mother's return, she did a lot of midwifery

A young Heidi (lower left) with her family in Kentucky

and enjoyed it immensely. Some people lived so far back in the woods that my Mom would ride horseback to help them. Mine was a childhood that was very much open and in touch with nature. I was always outside playing with the neighbor kids. When it was time for me to go to school, my parents did not like the nearby public schools, so they sent me to a church school a half hour away in Lexington, Kentucky, where I stayed with some people they knew. I do not really remember anything about this except that I hated it and it didn't last very long.

At that point, my parents felt pressured to move to a place with better schools and more financial opportunity. They had initially gone to Appalachia because they wanted to be in the place of greatest need. People were so poor that my Dad's practice was pretty much based on the barter system, and cash was scarce. Because we could not get everything we needed by barter, it was not a sustainable situation. When I was eight, we moved to suburban Boston, and my Dad set up a suburban medical practice in the New England Memorial Hospital in Stoneham, Massachusetts. A downside of this move was that my mom could no longer practice midwifery without going back to school for three years, which would have been difficult for her at this stage. She had small children and lots of responsibilities and worked full-time as a registered nurse.

Some of my early formative memories, I think, were of Mom being a very independent and self-possessed person. She had her own expertise, and she gave us children a lot of freedom and independence. As time went by, I began to realize that women were not treated as equals within our church, and that grated on me. Observing the lesser role of women in our religion was formative influence in my early life. School was never difficult for me; however, even though I was first in the class in geometry, I was not encouraged to go on to trigonometry and calculus. This didn't surprise me at the time, but later on, when I missed the mathematical skills, I realized that this is a part of our culture: women aren't encouraged to excel in math. Thirty years later, I am not sure this has changed, but I hope so. In college in the early '70s, the generational movements of feminism, civil rights, and pacifism influenced me.

Many of my early jobs had to do with medicine and the hospital. I was a hospital janitor and a nurse's aide. Coming from a very medical family, I thought that I would probably become a doctor; that was an early and long-held ambition of mine. In high school, I fell in love with languages and in college decided that I would major in them. I spent my sophomore college year in France, learning French and getting my major credits out of the way. I realized that I wanted an activity that would be practical as I put myself through medical school, so in my last two years of college I received an associate's degree in nursing. I enjoyed the nursing-level sciences, and I very much enjoyed the practice of nursing, except for the fact that in the suburban hospital where I was trained, the older, traditional doctors didn't treat the nurses with much respect. Soon after I finished my nursing degree, one of my first full-time jobs was at Boston Children's Hospital. There I worked with residents and interns from Harvard Medical School and found the interactions and the respect for nurses to be totally different. They were engaged, they were very willing to teach, and they were actively interested in being partners in the care of sick children.

I decided to take a year off to return to Europe to work on my languages. During that year abroad, I traveled, and when I was short of money, it was easy to find temporary jobs for a month or two. Some of those jobs were in the French part of Switzerland and in the southern part of France. I got to know a lot of different people and enjoyed that independence.

I met my husband-to-be while visiting Florence, where common friends introduced us. Once it became clear that this was a serious relationship, I looked around for a nursing job there and found one very easily in a small private hospital run by Irish nuns. Already fluent in French, I learned Italian very easily, and with the nuns' help, my Italian became passable in a couple of months. At this point I found, to my amazement, that I would be eligible to attend Florence University Medical School (for only three hundred dollars a year!), so instead of going back to the States for medical school, I stayed in Florence and did the first two years of Italian medical school. It was quite an experience, taking classes with two hundred fifty other first-year medical students in a giant hall built a few centuries ago, trying to understand the lectures and take notes (in Italian) at the same time. I just loved the Italian language, and this was a wonderful experience because I was learning both medicine and the language at the same time.

After my second year in the Italian medical school, I learned that it was going to take three thousand dollars to have my American college transcripts translated through the American Embassy if I were to continue medical school in Italy. I didn't have that kind of money, and I figured if it were going to cost that much to continue in Italy, it would be worth it to go to medical school in the States. I took my MCAT in Rome and began applying to medical schools in the States. At this point, fate completely changed my plans. The different schools in which I was interested sent postcards asking for additional information to complete my packet, but they sent the postcards by regular mail; they didn't notice my address was in Italy! These postcards came by boat and got to me three months late, making me way past application deadlines for American medical schools. At this point I made a radical decision; because I really loved the pre-med basic sciences, I would apply to graduate school! My fiancé had just finished the Italian University with a laureate in mathematics. He really wanted to get an American PhD, so he applied and obtained a scholarship to an American university. We applied to the University of Texas at Austin (an Italian professor suggested he work with someone in the Math Department there) and off we went to graduate school together.

From the beginning, I loved graduate school; I loved the intensity and commitment and loved the discovery process. I never again thought about going to medical school—an interesting accident of fate, I guess, that has taken me into research instead of medicine. Graduate school was great fun. I found a major professor, Mike Menaker, who was from a Viennese intellectual family (his mom was a psychoanalyst who studied with Freud), and his fertile mind came up with at least one thesis project a day for me. I learned early to be independent, to smell out a good project, and then to push it hard and systematically. Within three years, I had a graduate thesis on the workings of the circadian clock in the pineal gland and retina of birds, which began my lifelong interest in vision and visual transduction.

From that time on, my career has been much more of a straight arrow. In quick succession, I got a postdoctoral fellowship and then a faculty position, launched my lab, received a grant, learned about mentoring students and technicians, and started to publish and to make a mark for myself. Living through it at the time, it seemed like a random walk; although the journey was always related to a medical career, fate directed particular decisions along the way. Looking back, I realize there was more directionality in that once I found something that I really loved, that was it. That was my career. It led me to a fellowship at the University of Wisconsin to study the biochemistry of visual transduction with Deric Bownds, George Wald's first graduate student, and a great photoreceptor biochemist. Soon after came a faculty position at the University of Illinois-Chicago in the Department of Physiology and Biophysics. There, I made my mark on the world of G protein-mediated signal transduction, where I was the first to view the high-resolution three-dimensional structure of a heterotrimeric G protein. With this fundamental discovery, my career has developed as I learned more and more about the mechanisms by which hormones and neurotransmitters have their influence on cellular processes.

Recalling my time as an assistant professor, I realize that one source of frustration came from not understanding the rules of the game. I did not grasp the importance of strong mentoring relation-ships, and there was nobody who guided me or told me that this was important. So I bumbled along, not as efficiently as I could have with a bit of mentoring. There were no dramatic difficulties as I went along that path; it was just not a natural thing to make easy alliances with my seniors, who were mostly somewhat older men. It is important for young scientists to have mentors with whom they feel comfortable. I do not feel, however, that I experienced any overt sexism. Ironically, the only time when I felt there was some unfairness in my career came from an interaction with a senior woman who was pretty hard on me. She was rather hostile, and I suspect that she may have thought that because she had made it in a man's world at a time when that was very unusual, there was no reason to make it easier on younger women coming after her. We never had the conversation, but I felt at the time it was ironic that the person who treated me somewhat harshly was a senior woman in the field.

Despite this one negative experience, a large number of men and women have been actively helpful in promoting me in ways that surprised me throughout my career. They helped me enormously in developing my career, gaining recognition, and being able to continue as a scientist. Some of this help came from sources I will never know.

What about the impact of a scientific career on my ability to have a family? There are many outstanding women scientists who do have children. It's amazing to me how often the question is posed to me: "Can women have a career and family, or any other life outside of a career?" Not only can women have a "life" along with their career, but many have done it successfully and are really happy with the results.

I have been tremendously lucky in two things. First is having a husband who is highly supportive of my role as a scientist, and second is having a similar commitment to our work. There are challenges in a dual-faculty family—whenever either of us was not in an optimal situation, we both had to be willing to make compromises along the way. Those compromises were nothing in comparison with the positives of having a devoted life partner who gives and takes good advice, who has the same commitment to science, who is willing to work long hours, and who understands both the joy of discovery and the frustration when things don't work.

Neither of us cared about having children. Our careers progressed at similar paces, so we were very actively engaged in pursuing them all through the years that would have been our child-rearing years. We had very dynamic lives and were able to travel whenever we wanted. So I don't have the constraints that women have in trying to juggle both a career and children. On the other hand, I am the principal caregiver to my elderly parents!

As my tenure came through and my career advanced, I had a life-shaping experience at a Cold Spring Harbor meeting on signal transduction. It was a very big and influential meeting with two hundred fifty speakers and only one of them a woman. There were a lot of dynamic women in the field, and their absence on the program was commented on widely. At this time I was feeling more secure in my own career, so I decided to do something about this situation. Even though there were lots of women scientists making successful discoveries, they did not necessarily get invited to speak at platform sessions. So at that point I started to become more active in organizing meetings to ensure that prominent scientists, younger and female as well as male, were invited. I continue to be an activist in my scientific societies, and one of my goals is to make sure that women and minorities are at the table, able to attend the scientific meetings and to be actively involved in the societies. I am president-elect of the American Society for Biochemistry and Molecular Biology. During my term as president, I hope to be a strong advocate of support for biomedical research in this country. Broad investments in science have been an engine for better health and longer life, and we are in a period of incredible discovery. A flat or decreasing federal research budget will erode the catalytic effects of the recent doubling of the National Institutes of Health (NIH) budget on medical research breakthroughs. Hardest hit in this tight funding environment will be young people who have been superbly trained and whose contributions could continue for many years. If funding levels do not improve, we will lose these young minds to other careers, and the tremendous investment the scientific community has made in their training will be lost.

Some fifteen years into my career, I began to be asked to apply for department chair positions. I was at first very hesitant but decided to accept the position at Vanderbilt because of the honor associated with the Department of Pharmacology, one of the leaders in the country by reputation, citation analysis, and NIH funding. In my first four years as chair, I have overseen an increase of the size of the department and a doubling in its NIH funding. The department's strengths lie in G protein-related signal transduction and neuroscience, and I have expanded the areas of structural biology of membrane proteins.

One passion I have developed is for turning knowledge of G protein mechanisms into strategies to alter and modify signaling processes when they go awry. My work has revealed the details of G protein coupled receptors (GPCR) interactions with G proteins, and I've been pushing to find ways to interrupt this interface with small molecules.

This is a very rich area: GPCRs make up one of the largest superfamilies with more than three hundred pharmaceutically relevant members and there are drugs on the market for only 5 percent of them. Even so, 50 percent of all drugs target these receptors. My work is the first to show that it's possible to target the interaction in a specific way. Recently, my laboratory has begun to identify systems functions of G protein subunits within the context of integrated physiological responses, and we are applying mathematical modeling approaches to understand these networks of G protein signaling pathways. In the future, these integrative modeling approaches will allow us to predict excellent targets within signaling cascades for novel therapeutics.

As a piece of advice for young women scientists, I would urge you to seek out champions and ask them to nominate you for speaking engagements, awards, and other recognitions. The fact that women may not be nominated for these awards as frequently as men may be a result of subtle sexism. You may also encounter male scientists who assume they will be better competitors than their female colleagues. Science is a communal activity; progress comes both from singular discoveries (very rare) as well as building on what has come before. You need to be scientifically aggressive and a proponent of your discoveries. None of this is easy, and we are all worse at this than we should be. You also need to develop self-knowledge, and a part of this is to understand that our socialization as women has built-in brakes to assertiveness in many different ways. I believe that this is one of the most important reasons women scientists are not as recognized as they deserve to be.

Chapter 23:
NEERAJA B. PETERSON, MD, MSc
Assistant Professor of Medicine

I joined the Division of General Internal Medicine and Public Health at Vanderbilt in 2002. I conduct health services research; my research focus is in cancer prevention and women's health. I am the recipient of a BIRCWH (Building Interdisciplinary Research Careers in Women's Health) career development award, the National Institutes of Health's (NIH) Loan Repayment Program, and intramural funding at Vanderbilt. I also have an active primary care practice in Vanderbilt's Adult Primary Care Clinic (APCC). My teaching responsibilities include instructing internal medicine residents in their continuity clinics weekly and attending on the general internal medicine wards one month a year.

I was born in Starkville, Mississippi, and lived my childhood in Tupelo (the birthplace of Elvis, our only claim to fame). My parents came to the United States from India in 1969 so that my dad could complete his master's degree in industrial and mechanical engineering at Mississippi State. Being a member of the only Indian family in Tupelo, I felt at times completely out of place but most other times happy and content. My mother stayed home when I was little and became a teacher at Head Start when I began elementary school. I was very close to my older brother, who was only a year older than me. My first real exposure to the medical profession came as a volunteer (a candy striper) when I was fifteen. I ended up working in the hospital gift shop, initially as a volunteer and later as a part-time employee in the evenings and on weekends. I met a lot of interesting people in the gift shop but was relatively insulated from true hospital life.

Neeraja Peterson and her husband, Josh, with their children

I did very well in school and was one of six valedictorians—all female! I was one of a handful of graduates from my high school that chose to go out of state for college. At Duke University, I majored in biomedical engineering. I did not want to be like every other Indian kid who wanted to be a doctor, and I admired my dad's engineering profession so biomedical engineering seemed a good choice. Unfortunately, in my junior year, I decided that I was not cut out for engineering. I thought, "Oh, no, what am I going to do now?" My brother, having majored in chemical engineering, had decided to go medical school and convinced me that I, too, should apply to medical school. I went to summer school to complete my pre-med classes, took the MCATs (got a decent but not fantastic score), and applied to medical school. Again, my brother intervened and convinced me to come to Vanderbilt, where he was in medical school. We lived together for three years in medical school, and I met my future husband during my first year. I was exposed to basic science research during my second year (a required course). I enjoyed the laboratory experience but did not think I was cut out for a career at the bench. I wanted the satisfaction of direct patient care, so I set off on the path of a clinical career.

My husband and I couples-matched in internal medicine at Duke University. After completing a tough but rewarding residency, we were ready to move on to the next stage of our careers. My husband was interested in clinical research and accepted a fellowship position in general internal medicine at the Brigham and Women's Hospital in Boston. We knew that we would likely be in Boston only for two years while he completed his fellowship, so I again thought, "Oh, no, what am I going to do now?" I interviewed for a couple of private practice jobs but was very discouraged when my interviews focused on the "bottom line" and how to meet my financial benchmark. As a resident, I was not used to thinking about how many patients I needed to see to get paid and not get kicked out of my practice. I spoke to some mentors at Duke. One mentor who ran the Women's Health Center at the Durham Veterans Administration Hospital suggested that I consider a women's health research fellowship. She put me in touch with her colleague at Boston University. Fortunately, they had a position available, and I took it. I did not really know what

I was getting myself into, but knew I was avoiding the big, bad world of private practice.

Fellowship was a wonderful experience, both professionally and personally. When I started fellowship, I knew next to nothing about epidemiology, biostatistics, and health services research, but that quickly changed. As part of my fellowship, I completed a master's of science in epidemiology. In the two years I was in Boston, I took classes; learned more about women's health; completed two fellowship research projects; discovered my career home in academic medicine; had my first son; and became pregnant with my second. Fellowship taught me the importance of mentorship. I was fortunate to work with two mentors, one male and one female. From each I learned valuable lessons about organizing research projects, writing grants and manuscripts, and juggling family and career. One of my mentors was Robert Friedman, professor of Medicine, director of the Boston University (BU) Medical Information Systems Unit, and director of the General Internal Medicine Fellowship Program at BU. My other mentor was Karen Freund, professor of Medicine and director of the BU Center of Excellence in Women's Health.

My husband and I accepted positions as physician-scientists at Vanderbilt in 2002. I remember feeling nervous about telling Bob Dittus, the division chief of General Internal Medicine and Public Health at Vanderbilt, about my second pregnancy and that I would not be able to start in July like the other junior faculty. I wanted to start a few months later after having my son. I was pleasantly surprised that he was so supportive, and continued to be so, not only of me but also of the entire female faculty in our division with young children. As one example, my office at Vanderbilt had glass windows from which you could see inside from the hallway. I planned on pumping breastmilk at work so that I could breastfeed my son for six months. However, this would have been difficult because everyone could see into my office. When I told Bob's secretary about my concern, she said they could order blinds for my windows. In fact, they had done the same thing for two other women faculty so that they could also pump at work.

As I have progressed in my career, mentorship has played a vital role in all stages of my career so far—from being a fellow to junior faculty, and, I suspect, will continue to be important as I advance in my career. I have been fortunate to have good mentorship at Vanderbilt. I do not have one person that I consider my primary mentor, but instead three senior mentors and a group of peer-to-peer mentors (junior faculty colleagues). I also receive informal mentorship from many others as the need arises; the collegial atmosphere at Vanderbilt makes this possible. One of my primary mentors is Dittus, who, although very busy with the many hats he wears as chief of my division, is always available, from helping me edit a paper, to generating ideas for grant proposals, to offering advice about buying a car. My other mentors at Vanderbilt include Kathleen Egan, associate professor of Medicine, and Kathleen Dwyer, associate professor of Nursing.

My husband and I both work full-time. We could not do this without excellent child care. We have a full-time nanny who takes our boys to preschool two days a week and keeps them well entertained the rest of the week. She can stay home with them when they are sick, and she runs errands while the boys are in preschool. We also have someone who comes once every two weeks to clean our house.

I am in my office by 8 A.M. most mornings and am home by 5:30 P.M. I rarely do job-related work at home. When I get home, I spend the time with my kids. After they go to bed, I help clean up after dinner and spend about an hour relaxing before going to bed. Unfortunately, I am not one of those people that can survive on only five hours of sleep; I need a full seven or eight. I will occasionally work in the evenings, usually before a grant deadline or a self-imposed deadline of some sort. I try to be very productive when I am at work. This often means working through lunch and eating at my desk.

In a perfect world, I would love to be with my children more. However, I love my job and believe that it offers me the ability to spend more time with my boys than would a private practice. With the type of research that I do, my work schedule is very predictable, and I am home most days at a reasonable time. On the days that I am not, it usually means that I have gone to the gym to exercise after work. When one of my boys has a doctor's appointment or a special event at preschool, my flexible schedule usually allows me to attend it. My call group is large, so that I am only on call for my practice four weekends a year. That leaves me all the other weekends (except for the one month I am rounding on the inpatient wards) free at home with my children. I never have to go into the hospital at night for clinical duties, as our practice is well covered with moonlighting residents and fellows.

I think one of the reasons I enjoy seeing patients so much in clinic is because I do not have to do it every day. Patient care is very rewarding, but it can also be overwhelming when done 24/7. Health services research allows me the opportunity to pursue intellectually stimulating work at a less frenetic pace. I can discuss research ideas with colleagues, write papers and grants, and analyze data in my office. And I feel that I am contributing to medical knowledge, which I hope will lead to improved patient care one day.

I would encourage all women interested in research careers to weigh the pros and cons carefully. I think you will find that the pros quickly outweigh the cons. Some women may worry about the impact of research careers on financial stability and family life. I would argue that it can be financially rewarding. I was fortunate to become a recipient of the Loan Repayment Program through the NIH. This two-year program pays a substantial portion of a new investigator's loans and can be competitively renewed. I would also argue that, with its predictable and flexible scheduling, a research career is very good for family life.

Chapter 24:
XIAO-OU SHU, MD, MPH, PHD
Professor of Medicine and Pediatrics

I am a professor of Medicine in the Division of General Internal Medicine and Public Health in the Department of Medicine, where I study the epidemiology of cancer and other chronic diseases. I hold five National Institutes of Health (NIH) and Department of Defense grants with annual research funding of approximately $1.7 million. Over the years, I have studied the epidemiology of childhood cancers, pregnancy outcomes, adult cancers, and chronic diseases of women. I have more than two hundred publications in peer-reviewed journals and enjoy the challenge of research and the rewards of seeing research results translated into preventive observations.

Fifteen years ago, I stepped off a plane from China and into the largest and busiest city in the United States with a scholarship to Columbia University in New York City. I had arrived ready to pursue graduate studies in epidemiology but not before serious debate ensued over whether I should practice medicine or pursue research.

I had grown up in China surrounded by a rich and diverse legacy of doctors. My paternal grandfather had practiced traditional Chinese medicine. A kindly old man with a long white beard, he ran a pharmacy with long walls of small drawers filled with herbs and was well respected by the people in his small village, whom he treated with herbs and acupuncture. When I last saw him, I was five years old. Probably suffering from a neurological disease, his hand shook so uncontrollably at the dinner table that he was

The Shu family vacationing at Monticello

151

hardly able to pick up any food with his slim chopsticks. In the end, he was unable to cure his own disease and died three years later, which is when I realized that such traditional medicine has its limitations.

In contrast, my maternal grandfather came from a wealthy family and received a formal training in Western medicine from a prestigious medical school. Through an arranged marriage, he wed a beautiful girl from a small town—my grandmother. My mother was their oldest child. My grandfather, however, was often absent and did not love my grandmother, who had no formal education. He eventually left her and their children at the family compound and married one of his nurses. Unfortunately, both he and his second wife came down with tuberculosis and died a couple of years later.

My mother was twelve when her father died. After seeing the suffering of her uneducated mother, she was determined to get a good education and follow her father into the world of medicine. Her wealthy uncles provided a good standard of living, but they refused to support my mother's dream of becoming a doctor because, like most men at the time, they were heavily influenced by the Confucian philosophy that "the woman with no talent is the one who has merit." Unhappy with her comfortable but stilted lifestyle, my mother ran away at the age of fourteen to attend a free boarding school for teachers, the only option for education available to her.

Though she never became a doctor, my mother did become the first female teacher in her county and instilled her dreams in my sisters and me. Growing up hearing stories of our grandfathers, we, too, wanted to join the world of heroic white-clad doctors. Everything changed, however, in 1966 when the Cultural Revolution began in China. Most universities no longer admitted students based on academic merits, and students like my two older sisters were removed from high schools or universities and sent off for "re-education" as farmers and workers. For my sisters and me, the door to higher education was doubly closed because we grew up in a comfortable city life and my parents came from wealthy families, a class of youth that was particularly targeted for re-education.

At the age of sixteen, I was sent to the countryside for re-education as my sisters had been. Neither of them ever had the chance to go back and finish school, but I got lucky. One year after I left home, the Cultural Revolution ended, and I was able to pursue the family dream of entering medicine. Facing loads of work I had missed, I struggled for two years to catch up before finally passing the very competitive university entrance exams. The day I received my acceptance letter from Shanghai Medical University was one of the happiest days of my life. I was finally realizing both my dream and my mother's.

However, not long after that I felt like I had been dropped from a skyscraper when I learned that I had been assigned to major in preventive medicine, not clinical medicine as I had expected. I will always remember the first day we met our professors when, still disappointed, I sat with a crowd of other students in a big lecture room. As we shuffled our papers, a professor, one of the most prominent epidemiologists in the country, told us, "The best doctor is the one who treats people before they get a disease." At the time, I thought that he was only comforting disappointed students like me. I was determined to switch to clinical practice after completing my medical education.

It was not until my internship in my fourth year that my desire to practice medicine started to change. Sitting in a hot, crowded clinic one day, I realized I was seeing patients over and over again with the same avoidable diseases, such as gastritis, hypertension, and diabetes. Frustrated with having constantly to treat these conditions, I remembered what my professor had told us on that first day and started thinking how much better it would be to prevent these diseases instead of treating them again and again.

During my final year of medical training, and with the idea of switching majors still lurking in my mind, I took courses such as epidemiology, biostatistics, environmental health, and childhood nutrition and development. After learning the stories behind how the cause of cholera was found or that cigarette smoking was linked to lung cancer, I was hooked on epidemiology. Epidemiology appealed to the part of me that had been so frustrated by my experience in the clinic; out of its focus on finding causative answers, doctors might prevent patients from getting sick in the first place. I also found that I loved the detective nature of epidemiology; everything was a puzzle with solutions to be found only by good observations, critical analysis, and well-planned experiments.

After receiving my medical degree, I entered a graduate program in public health and majored in cancer epidemiology. I conducted my first epidemiological study on childhood leukemia and had the opportunity to visit the National Cancer Institute (NCI) in Maryland. Working with several world-class cancer epidemiologists at NCI, I published my first paper in the journal *Lancet*, linking chlorampenica use, a drug commonly used to treat infectious disease in developing countries at that time, to risk of childhood leukemia. This exciting experience was the spark that started my research career, and three years after obtaining my master's degree in public health, I arrived in New York to start on a doctoral degree in epidemiology. I graduated receiving the Anna C. Celman Award for Excellence in Epidemiology.

The experience of having lived in two countries with immensely different lifestyles and cultures gave me a deep appreciation of how those differences could affect health. Witnessing a rapid increase of cancers, diabetes, and cardiovascular diseases among the Chinese as their lifestyles became Westernized led me to investigate the factors that are part of a traditional Chinese lifestyle—such factors as dietary soy intake, ginseng use, and Tai Chi, commonly used in China to help prevent these Western diseases. My research has naturally extended to the investigation of how genetic susceptibility contributes to cancer and other chronic diseases.

My research has been very rewarding for me both professionally and personally. Projects I have been involved with have found, for example, that adolescent soy consumption may reduce the risk of breast cancer in later life, and that soy food consumption in women may reduce the risk of coronary heart disease as well as bone fractures in postmenopausal women. In addition, my research has shown that ginseng use may improve survival and quality of life for breast cancer patients, and that polymorphisms in the cyclin D1, TGFβ, VEGF, and PAI-1 genes may influence breast cancer survival. These findings are satisfying professionally in that they are likely to have significant impact on the prevention of these diseases. But they are also satisfying in that they provide a link back to the wisdom of my paternal grandfather and the values of traditional Chinese medicine.

Perhaps even more important, I am happily married with two lovely children. While it is always a challenge to balance a career and one's family life, I have been able to carry out my research away from the office from time to time—as many epidemiologists can—with the help of technology. This relatively flexible working schedule has allowed me to balance the needs of my children and family with the demands of my research. At the same time, my career has benefited from the strong support of my family. I have been very fortunate that my husband, also a professor of epidemiology, has been at my side to lend wisdom and professional and emotional support throughout my career. I am also very grateful to my parents, parent-in-law, and my children for their love and support. It is my hope that my children will carry on my legacy to become doctors who "treat people before they get disease."

Chapter 25:
KATHERINE T. MURRAY, MD
Associate Professor of Medicine and Pharmacology

As a physician-scientist at Vanderbilt, I believe that I have the best of both worlds. I am currently associate professor of Medicine and Pharmacology, with appointments in the divisions of Clinical Pharmacology and Cardiology. Given my clinical training in cardiac electrophysiology, the focus of my practice is the management of patients with cardiac arrhythmias. I attend on the inpatient cardiac arrhythmia service for two one-half month blocks per year, with a half-day clinic each week. The majority of my time is spent supervising basic research efforts to investigate the molecular basis of cardiac electrophysiology and arrhythmia susceptibility, with an overall goal to develop novel therapeutic approaches to treating cardiac arrhythmias. I am currently principal investigator on two grants from the National Institutes of Health (NIH) and co-investigator on three additional collaborative NIH grants, with additional funding from the American Heart Association and pharmaceutical industry-based grants. As an established investigator, I participate in the peer-review process for NIH and American Heart Association grants, manuscripts submitted to multiple journals, and abstracts submitted for presentation at major international scientific meetings. I have also served on the arrhythmia and mortality event committees for industry- and NIH-based clinical trials. An important component of my activities have been the supervision of numerous (more than twenty) trainees in my laboratory, who have included undergraduate, graduate, and

Kathy Murray in her laboratory

medical students, post-docs, and junior faculty. I am involved in teaching the principles of basic and clinical cardiovascular electro-physiology at many venues. I lecture each year in the medical school pharmacology course, an advanced cardiovascular pharmacology course, and undergraduate engineering classes, in addition to teaching arrhythmia management to medical students, house staff, and cardi-ology and electrophysiology fellows. I also participate in educational seminars/tutorials at major meetings.

As I grew up, I was naturally attracted to science and math-related subjects during school. My father was an orthopedic surgeon, and I was fascinated by his tales from the hospital, as well as his horrific slides of surgical procedures that would be interspersed among those of family vacations. During the summers in high school, I worked in a hospital as a medical technician, performing miscellaneous tasks such as drawing blood and obtaining electrocardiograms. I will never forget witnessing my first code (a patient in cardiac arrest): it was terrifying and yet thrilling, and I was struck by the enormity of the physician's role during the arrest. In retrospect, I am certain that this experience initiated my journey to become a doctor.

My strong interest in science and medicine drew me to Duke Uni-versity for college (as a zoology major) and ultimately medical school. The unusual curriculum format at Duke Medical School permitted a year of research during the third year. Under the tutelage of a terrific mentor, Gerald Serwer in Pediatric Cardiology, I conducted a study to evaluate the use of echocardiography to detect hemodynamic complications following surgical repair of transposition of the great vessels (a congenital heart disease). The results were presented at a cardio-vascular research forum at Duke, and in a first-author paper published in the journal *Circulation*. It was a wonderful year that generated an intense interest in cardiology and a zeal for investigation that has never left me.

During residency, I was hesitant to commit to a cardiology fellow-ship, believing that this was not compatible with having a family. I spent the next two years as a fellow in clinical pharmacology working with the arrhythmia service at Vanderbilt. This experience afforded me the opportunity to work with Raymond Woosley and Dan Roden, to conduct Clinical Research Center–based patient studies, to gain expertise in determining antiarrhythmic drug pharmacokinetics in humans, to conduct rounds on the arrhythmia inpatient service on a daily basis, to develop assays of drugs and their metabolites, and to learn basic techniques for recording action potentials from cardiac tissue to determine the electrophysiologic effects of drugs. The experience was very productive and enormously satisfying, and I became firmly committed to a career as an academic clinical and basic electrophysiologist.

After clinical training in cardiology and clinical electrophysiology at Duke, I returned to Vanderbilt and spent a year learning patch clamp techniques for recording individual ionic currents. Subsequently, I joined the faculty, obtained extramural funding, and ultimately started my own laboratory. The importance of proper mentoring to this point in my career cannot be overstated—it was critical to my success as I explored different research venues and vital for my path ultimately to academic independence.

As I grew up, my parents were always there for me, and I have strived very hard to do the same for my own children. They are currently teenagers in the throes of adolescence, a time when ever-present parental interaction is never more important. While I work hard and typically long hours, my schedule allows great flexibility so that I do not miss important events at school or on the athletic field. By necessity, I must be very efficient with my time, and I focus my efforts on what I really need or want to do. Thus, an important element of time management is setting priorities and learning to say no; I simply can't do everything I am asked to do and at the same time be an effective mother, physician, and researcher.

To make it all work, I am blessed with a wonderful physician husband who not only enjoys being with the children as much as I do, but who is also very supportive of what I do professionally. When I am on call (typically a twelve-to-fourteen-hour-a-day job on the arrhythmia service), he is Mr. Mom. While I am usually the first parent home at the end of the day, he will fill in for me if needed, such as during arrhythmia call or at grant submission time. There is a well-defined division of labor once we all get home. Our vacations are always family ones.

Another critical element for us has been the continual presence of a full-time nanny during the day. We have made sure that this person is both loving and highly responsible. We have preferred in-home child care for our children, as this was what we enjoyed growing up. Our nanny handles transportation for after-school activities, errands, preparing simple dinners, laundry, grocery shopping, and midday care for the dogs. It is a very rare event for her to work past 6 P.M. We have a separate person who cleans the house weekly.

I thoroughly enjoy both the clinical and basic aspects of my job. Treating arrhythmias is exciting—we often meet our patients during or immediately following a cardiac arrest. It is a very rewarding specialty and immediately gratifying; helping an individual get back in order his/her life medically and psychosocially after such an event is no small task, and a close physician-patient relationship naturally results. With respect to basic research, I sometimes can't believe I get paid to do this. The constant evolution of knowledge and experimental techniques that occurs at the cutting edge of important scientific questions provides tremendous intellectual excitement. If you enjoy constantly learning new things, this is a wonderful existence! The translational aspects of exploring questions from the bedside to the bench, and back again, provide a more delayed but equally satisfying type of gratification, with the anticipation that large numbers of patients may ultimately benefit. Scientific progress is achieved in concert with others in the laboratory, and often in collaboration with those in other labs, fostering an atmosphere of collegiality. Importantly, it is a privilege to participate in the maturation of young scientists as they morph towards becoming independent investigators—this is definitely one of the most enjoyable aspects of the job. However, a word of caution—tenacity (or stubbornness) is an important trait too for a happy life in basic research. Hypotheses are proven wrong, and papers and grants are rejected, as a natural part of the process; hence, "alligator skin" is definitely a plus.

Chapter 26:
AGNES B. FOGO, MD
Professor of Pathology, Medicine, and Pediatrics

I am a renal pathologist at Vanderbilt and direct the Renal/Electron Microscopy Laboratory, which is responsible for diagnostic renal biopsy work both here and from a large and active outside referral practice. I am very fortunate that my research is tightly linked to my clinical work. Our research laboratory works on progression of chronic kidney disease, using animal models and cell cultures to understand mechanisms of scarring and its possible regression. In addition, I am privileged to teach, particularly second-year medical students in the pathology course; residents, fellows, and visiting scholars in our clinical laboratory; and students and postdoctoral fellows in our research laboratories. We have had a lot of fun with our focus on renal disease and are fortunate to be well-funded. The clinical practice is busy, including transplant biopsies and medical renal biopsies of children and adults with a wide variety of lesions.

I was born in New York to Norwegian parents who moved back to Norway when I was seven years old. I grew up bilingual and with firm roots both in America and in Norway. My younger sister is a doctor in internal medicine, and my kid brother is a bona fide computer nerd, working with computer support services at the University of Oslo. The sciences were always inter-esting to me. My teachers were inspirational and memorable, and they actually seemed to enjoy interacting with their students. In high school, I focused on science and was also involved in sports,

Agnes Fogo (left)
teaching renal pathology

student government, the theater group—and anything else I could get my hands on. It was during this time that I developed my abiding passion for basketball. Our club team had a lot of fun, and I still am in touch with friends from those days, laughing at our adventures, challenges, and accomplishments as a team.

As I neared the end of high school, I wasn't certain whether I wanted to pursue medicine or another branch of science but decided to give medicine a shot. In Norway, as in much of Europe, medical studies are a combined pre-medical and clinical curriculum lasting six and a half years. Even the preclinical curriculum to which we were exposed during our first year was enthralling. The intricacies of genetics, biochemistry, and biology totally captured my imagination. For personal reasons, I decided to move to the United States and continued my pre-medical requirements at the University of Tennessee at Chattanooga, where I majored in biology. I was very fortunate to have inspirational teachers there as well. Apart from deciding that I never wanted to pith another frog again, I thoroughly enjoyed my experiences as a teaching assistant in the physiology laboratory.

When I visited Vanderbilt School of Medicine as a prospective applicant, I was most impressed with the people I encountered. On an in-depth weekend visit as an applicant for a scholarship, I had the opportunity to meet Grant Liddle, Addison Scoville, Oscar Crofford, and other leaders of the medical school. The other students who also had been accepted to Vanderbilt were so accomplished, and the atmosphere was so embracing, collegial, and positive, I decided that this was the place for me. The scholarship I was fortunate enough to receive made it possible to follow through on this decision without financial hardship; it was truly a wonderful generous opportunity provided by the Potter, Wilson, and Menafee families.

Medical school proved to be everything I had envisioned and then some. While I would never wish to take biochemistry again, the intense learning happened in a very supportive environment. During the summers, I had special opportunities to pursue research, in genetics after my first summer and in pathology with Robert Collins after my second year. This further whetted my appetite for scientific work. The ability to see tissues and understand pathophysiologic processes was exciting. Collins made all his students think rigorously about mechanisms of disease. During the fourth year, I was able to conduct research with Sandy Lawton for a brief period and with Jacek Hawiger and Collins. These experiences further cemented my decision that pathology and academic medicine would make a good match.

It also became apparent to me during this time that the usual track of medical school followed by internship, residency, fellowship, and beginning academic appointment did not leave time for personal pursuits such as starting a family. The advice and support of my faculty mentors were very helpful as I decided to delay residency for a year while my husband and I planned to start a family. This was quite uncommon in those days, yet every single one of the faculty, including my prospective chair in Pathology, Bill Hartmann, Liddle in Medicine, and Collins, were supportive and kind. I will never forget the attention and kindness from Janelle Owens, assistant to Dean Chapman, as she planned for all eventualities so I could receive my diploma at graduation on May 15. Our first daughter cooperated beautifully and was born on May 6, so graduation was truly momentous. During this year of delay before starting pathology residency at Vanderbilt, I did a research associate fellowship, again with Hawiger and Collins. The laboratory environment was exciting and stimulating, and I learned from my wonderful mentors a lot about approaching research questions and data and dealing with failures.

Pathology residency was challenging, but it still seemed to be slightly easier than taking care of two toddlers on my off days (our second child arrived during my second year of residency). We had wonderful babysitters, who treated our children like an extension of their own families. My husband was working as an engineer at the time, and it was fortunate that his hours were more controlled than mine.

The opportunity to be involved in renal pathology came when I was a senior resident. Harry Jacobson and Iekuni Ichikawa had arrived at Vanderbilt, and they needed dedicated morphological expertise. Meanwhile, I had been trying to find a niche in anatomic pathology where I could develop a strong research interest. The opportunity to do a combined renal pathology clinical fellowship and research fellowship with Ichikawa and Alan Glick turned out to be a terrific match. Glick was a superb teacher with a great sense of humor and the most wonderful eye for morphology. It was a sad loss some years later when Glick died suddenly and unexpectedly of a myocardial infarction. He was one of my best friends and mentors in pathology, and his death left a void for many people at Vanderbilt.

Our third child was born during my chief residency/fellowship year. I can't say that I've ever been bored or wondered what to do with my free time since then! Our pathology chair was kind enough to offer me a tenure-track faculty position with protected time to develop research further. I continued to devote energies to learning more about diagnostic renal pathology and having fun in Ichikawa's lab, learning as much as I possibly could. I had the opportunity to be involved with teaching and the clinical service work, but still had significant protected time to develop expertise and experience in the laboratory, and most important, to try to write grant applications. Those first applications were overly ambitious and full of all types of beginner's errors of enthusiasm. The feedback and mentoring from Ichikawa was painful at times, but extraordinarily helpful and directed. I think I almost jumped through the ceiling for joy when my first grant was funded. These were truly thrilling years, seeing our children grow from babies to toddlers, and being part of a productive and vigorous research team, along with clinical work and teaching.

Of course, with all of these interests, there was never enough time for everything. My husband had changed jobs during this time and had more flexibility, so I no longer had to worry about drop off and pick up of children every day. I gave up tackling the mess in the house on an ongoing basis. For a brief time, we even extended ourselves to hire a housekeeper who did the most urgent cleaning work every couple of weeks. Although work in academia never ends, there often is a great deal of flexibility. Our children would take naps in the early afternoon, and probably had the latest bedtime of any toddlers around, so that we could spend some time together in the evening after I got home. Luckily, I am by nature a night owl and could work on important issues after they were in bed. It is amazing how much life can be simplified when you really focus and prioritize. One of our priorities has been to be involved with our children's sports activities. There certainly were times when it wasn't possible to get to a basketball game or soccer match, but most often it could be arranged. My children also always understood that if I could not be at such an event, it was because we were tied up with urgent patient care. They all knew how to spell "glomerulosclerosis" at a young age and have helped with typing references, copying papers, and taking phone messages. We all would work as a team to prepare for get-togethers at the house for the small student groups I would teach, and they certainly took pride in their contribution to pulling it all off. I have overheard my children almost bragging to their friends about how hard their mother works and the long hours that doctors keep. This is always followed by a shrug, to indicate that their mother truly is a hopeless case "but she loves her work."

The elements that make it fun to come in to work every day are the challenges of understanding a patient's disease and providing timely and important information for the nephrologist so that the patient can receive the best care. Equal excitement comes when an experiment works exactly as we had planned or—maybe more often—an unexpected result leads to new avenues of investigation and understanding.

The interactions with the students are particularly gratifying. Their talents and interest in learning and dedication are quite impressive. In addition, they are just fun young people to be around. Every year, they make me proud as they tackle and learn some of the challenging topics that we present in our pathology course. It is gratifying to receive a surprise e-mail or an indirect greeting from past students, recalling fondly their days at Vanderbilt. The progress and development of our fellows is another area of pride and joy. I am privileged to work with terrific people in the clinical and research laboratory. They all do a wonderful job and make it fun to tackle all of the challenges each day. Last but not least are the colleagues with whom I interact, both at Vanderbilt and at other medical centers as I attend meetings and conferences. To think out loud about ideas, experiments, and our various mutual interests with people like Iekuni, Tina Kon, or Nancy Brown is stimulating and invigorating.

I cannot imagine another field with more diverse opportunities to pursue those avenues that are most rewarding and exciting. In academic medicine, value is placed on discovery and mentoring and teaching students, as well as on the concrete goals of excellence in patient care, allowing for a very rich environment. I think it is important for young people, particularly women who may have specific time pressures related to family plans, to focus on and prioritize their goals. It is so easy to look at the career paths, schedules, and priorities of others and feel limited, even overwhelmed. In my experience, our leaders have great flexibility in trying to accommodate our various needs and allow us to develop our potential and talents fully.

I am most honored to contribute to this series of vignettes, and I hope that the wide spectrum of experiences and insights will inspire young potential scientists to allow us to mentor and help them achieve their goals in the way that works best for them.

Chapter 27:
LOUISE A. MAWN, MD
Assistant Professor of Ophthalmology

I am an assistant professor on the tenure track in the Department of Ophthalmology and full-time orbital surgeon, with a secondary appointment in Neurological Surgery. I do research. Equally important, I am a wife and mother.

As the only full-time ophthalmic plastic and reconstructive surgeon at Vanderbilt, I manage a large surgical practice. My educational background includes fellowship training in both ophthalmic plastic and reconstructive surgery and neuro-ophthalmology; consequently, my practice includes complex orbital pathology. I also maintain a very productive academic life, which focuses on advancing treatment of orbital disease. My overlapping training as both an orbital surgeon and neuro-ophthalmologist makes me uniquely qualified to address disease of the optic nerve from a surgical perspective. My grants from the National Institutes of Health (NIH) have focused on endoscopic methods for treating optic neuropathy and to create a real time intra-operative navigational system for the endoscope. I am also interested in applying neuropro-

tective agents directly to the optic nerve with the endoscope. I have been a member of a NIH study section to review grant proposals and hold a patent for my development of an orbital surgery instrument.

My commitments to professional organizations have led to my position as a board member in the American Society of Ophthalmic Plastic and

Louise Mawn with her children at Christmas

Reconstructive Surgery, my surgical subspecialty society, as well as other medical professional societies. To date, I have published thirty-eight peer-reviewed articles, written educational chapters for publications of the American Academy of Ophthalmology, authored book chapters, and have been an associate examiner for the American Board of Ophthalmology oral examinations.

At Vanderbilt, I rotate with my ophthalmology colleagues on the general ophthalmology call schedule and because of my expertise in periocular reconstruction, I am also available 24/7 for ophthalmic plastic trauma at one of the largest Level 1 trauma centers in the United States. In addition to this clinical activity, I coordinate the educational programs in ophthalmic plastic surgery for medical students and residents and serve as a mentor to the fellows in ophthalmic plastic and reconstructive surgery. This role has allowed me to guide closely eleven Vanderbilt medical students as they applied to ophthalmology residency programs, assist four residents obtain additional training in ophthalmic plastic surgery, and impact the training of three fellows in ophthalmic plastic surgery.

The circumstances of my life combined to make my present situation not only possible, but also extremely rewarding. Born the second of seven children with a close extended Irish-Catholic family, I grew up in what might be called organized chaos. Because we were close in age, my siblings and I studied and played together. We were encouraged to verbalize our thoughts, have the courage of our own opinions, and defend our conclusions in a reasoned manner. My mother was a teacher, my father an attorney and businessman. My parents instilled in us a love of education and hard work and served as prime examples of the maxim, "Giving is better than receiving." The small town outside of Boston where I grew up was a bedroom community of academics and professionals. Their children, my public high school classmates, were some of the brightest individuals I have ever encountered throughout my educational process and career.

Perhaps because our household was largely female, the idea that any field of endeavor might be more suited to men or might be too demanding for women, was never entertained. Our dad, very accomplished himself, emphasized constantly the need to keep reaching and testing ourselves until we achieved our personal best, and I believe we have. Three of my siblings are attorneys. One is a very successful investment banker. Two are graduate-trained instructional designers. We are all living happy, productive lives and have produced eighteen, soon to be nineteen, well-adjusted children.

Events that may seem trivial in the scheme of one's life often make a big difference in shaping our thoughts and ambitions. For instance, as a high myope, I was exposed to ophthalmology at an early age and was fascinated by the colorful paintings my ophthalmologist had collected during his medical missionary trips to Haiti and displayed along the top of his examination lane. I was, however, disappointed to find that this kindly old gentleman was quite skeptical about the safety and value of contact lenses. Fortunately for a young teenager with thick eyeglasses, further research brought us to another physician with a more open mind. This experience taught me that professional opinions may differ and careful research is worthwhile, particularly regarding refractive options.

Another interesting experience occurred when an eighth-grade classmate's father, a noted cardiovascular surgeon at Massachusetts General Hospital, invited the class to observe an open-heart surgery and tour the surgical suite and hospital. There was a glass-viewing pyramid, which looked directly down on the operating table. The class filed by the observation window and then drifted away, either repulsed or disinterested. While my classmates ate, I was drawn back to the observation window to gaze at the compelling operation below. The image remains vivid in my mind to this day, and I often feel that it helped me to find my way to a place in the field of surgery. I went on to pursue a dual major in English and chemistry at Duke University and during the summer of my junior year, studied English literature at Oxford University while trying to decide which field suited me best. As senior year ended, I became convinced that I wanted to be a doctor. I spent a year as a laboratory technician in neuropharmacology at Duke, primarily performing high pressure liquid chromatography on blood samples obtained in human experimental trials, and then began my medical training at Wake Forest University. In medical school, I enjoyed interacting with the medicine attendings and the erudite process of postulating the most likely diagnosis among the differential. I only remember encountering one female surgeon during the third-year clinical rotations—she was the chief resident in General Surgery. Like a character from *The House of God*, the other house officers called her "Lips" because she always applied a fresh coat of bright red lipstick prior to rounds with the attendings. Late in my third year, I took an away rotation at Tufts-New England Medical Center and met the academic consultative neuro-ophthalmologist who best exemplified the way I envisioned practicing medicine. This rotation also allowed me the opportunity to spend extensive time with a female resident who guided my career path into ophthalmology.

Ophthalmology combined both the intellectual aspects I was drawn to in medicine with what I felt was my natural talent for surgery. This intense experience introduced me to the life of an academician in a tertiary care medical center. This impression was confirmed during my fourth year of medical school, when I spent two months in an ophthalmic pathology research laboratory—the potential to immediately and directly impact the knowledge base in the field was obvious. I was fortunate enough to secure training in ophthalmology at one of the largest and most prestigious ophthalmic training programs in the country, the University of Iowa Hospitals and Clinics. The dedication of the full-time faculty to advancing the science of ophthalmology furthered my academic interest. I chose to pursue fellowship education in both neuro-ophthlamology and ophthalmic plastic and reconstructive surgery with the career goal of academic medicine so that I, too, could contribute to both patient care and visual science. Advisors noted that life would be easier as a clinically focused ophthalmologist where I would be more likely to balance family and work successfully. I found this advice disturbing because nothing in my career to that point indicated that I was searching for an easy position, but rather my educational choices and decisions all highlighted my drive to excel.

My fellowship mentors developed my knowledge base in the two subspecialty areas. Thomas R. Hedges III exposed me to the complexities of the anterior and posterior visual pathways, and David Jordan, with relentless commitment, honed my orbital surgical skills. Both of these men combined busy academic careers with a devotion to their children and expected I would be able to do the same. These men are like fathers to me, and my loyalty to them runs deep.

I met my husband during my busy third-year medical school clinical rotations. Unlike so many of the men I had dated, he never questioned my zeal for medicine or felt conflicted with the concept of a working wife. As we dated during my fourth year, I conveyed to him my hopes of becoming a subspecialized ophthalmologist. We shared the match process and the move to Iowa. During my residency, he entered a PhD program in health systems management and finished all course work and comprehensive exams during this time. My fellowship years in particular were challenging as we lived apart for the better part of two years. He worked as a consultant, spending the work week at Johns Hopkins during my first year; and during the second year of fellowship, I lived in Canada while he commuted between Boston and Baltimore. He completed his PhD dissertation after we joined the Vanderbilt faculty with the support of his chair.

At age thirty-four, I was surprised to find conceiving a child was more difficult than we had anticipated. After two separate reproductive endocrinologist evaluations, we had our first child, and I am currently pregnant with our fourth. The educational training programs, as well as combining a family and demanding academic career, have not been easy. Much of my academic work is done at night, after my children are asleep, or on the weekends. Many of the trauma cases also surface as patients during these sleep hours, which minimally disrupts my family schedule. We make every effort to have a family dinner time and evening together.

My children have all attended the Vanderbilt on-site child care center since age six weeks and have frequently accompanied me to the hospital after hours and on weekends. They proudly note that they go to "Vanderbilt." My daughter questions the hours that I work and the demands of my career, but she is fiercely proud of the fact that I am a doctor who helps people. Periocular trauma call, in particular, frequently requires my presence in the operating room. My daughter at age five knew enough to ask what kind of a dog was involved in a trauma call (as she knew that dogs like "pit bulls" could potentially threaten her own Halloween activities). My son once asked if all doctors were mothers, and when I replied no, he sang out with joy that he was one of the lucky ones.

I also am one of the lucky ones—I have the unique opportunity of contributing to society in a rewarding and meaningful way and also the good fortune of a wonderful family. Additionally, my every leisure moment is devoted to my children. They travel with us to academic meetings and on any other trips we may take. They have devoted grandparents, aunts, uncles, and cousins whom they love and who love them.

As a tertiary care physician, I have the opportunity to impact individual lives significantly, whether by reconstructing a face that has been unrecognizably traumatized or taking out an orbital apex lesion that threatens vision or life. By passing this knowledge on to younger physicians, I can hand these skills down through time. Through my research, I will potentially influence the health of people that I will never meet. There is no question that the complex orbital pathology that I encounter in my practice drives my desire to find a better way to treat optic neuropathies. The intense motivation derived from knowing real patients whose lives would be different if better treatments were available helps sustain me through the additional hours this research requires. Mentoring medical students through their own educational and decision-making processes is one of the additional opportunities I have enjoyed, particularly guiding them through the process of choosing ophthalmology as a career and then assisting them as they have negotiated this pathway.

My advice to a young woman contemplating a career in medical investigation would be to make your mark—do not waste your intelligence and potential. Learn as much as you can about one area, and contemplate how you could advance the science in that field. You will get out of life what you put into it. Do not believe the people who say it cannot be done and that you cannot have both a full academic life and a family. Highly efficient people get the most done. Learn to do some good with your life, and you will have a good life. I love what I do everyday—you can, too.

Epilogue

ERIC G. NEILSON, MD

Hugh Jackson Morgan Professor of Medicine

The science in medicine has flowered in a golden era for nearly sixty years. Today, an extraordinary diversity of ideas covers the landscape of biomedical research. One might ask how this creative spirit came about. As a medical student I certainly wondered until I realized that physicians, among others, could also be scientists. A laboratory notebook is full of blank pages for anyone to write on—but who will do it? Goethe speaks of the drive to find things out as "Der Innere Geist," or the spirit within. The life of an investigator is emblematic of this spirit, and much has been written of the personal satisfaction that intertwines a life of discovery. For demographic and political reasons, historical citations of this creative spirit are often associated with men. While this does not make sense and is disquieting, the accomplishments of women will be increasingly recounted as the gender of those choosing biomedical careers achieves better balance. Women in growing numbers are matriculating into graduate studies or training for clinical medicine. If they choose a life in science, will they transform the customs and conventions governing older, mostly male investigators as we know them today? The answer, of course, is yes. The unique perspective of women will inevitably torque the world of science, and their assertiveness will happily add value.

This anthology was assembled in hopes of demystifying an inspiring reality—women do science and do it with great distinction. The stories in this book are reminiscences from one community that cross decades of experience; some of these individuals are just starting, many are in the thick of it, and a few just finishing their lifework. All are notable for having tried to make a difference. The observations emanating from their lives are reaffirming and hopeful. In all these authors one quickly recognizes an exceptional love of science, commitment to problem solving, pleasure taken from hard work, and the satisfaction of finding something new. Springing from humble beginnings, middle-class comfort, or genuine wealth, their unfolded experiences suggest an interest in science has no preordained phenotype.

What shines through for all is a refined sense of curiosity. This is a fundamental trait of the biomedical scientist; interest in knowing why things happen paces the internal clock of any serious investigator.

The biochemist Albert Szent-Gyorgi's famous quip that "discovery consists of seeing what everybody has seen and thinking what nobody has thought" cleverly articulates the critical essence of research. Working at science is serious business, and unraveling a biologic mystery can be fascinating stuff. It takes just as much effort, however, to answer a trifling question as it does a deep one, so the effort is far more pleasurable if the problems are worthy and stir imagination. Success and sustainability are not mutually exclusive. Choosing a problem of great personal interest and sticking with it for the pleasure of finding ensures sustainability. One senses in these reminiscences a special joy in working on questions of value.

Many of the stories in this anthology also touch on the art of facing adversity. None of us is immune to odd moments of gloom. T. H. Huxley, that great explainer of science, once suggested that glumness appears "with the slaying of a beautiful hypothesis by an ugly fact." Talented investigators who survive these flashes of insecurity have a strong inner compass that magically points forward when flagging experiments need resuscitation. Personalities with resiliency typically puzzle out new solutions to a problem without lasting self-doubt. As John Tarpley, Vanderbilt's resident wag, is fond of saying, "in life adversity is certain, misery is optional." Happiness, after all, is an inside job.

It is also clear that to succeed in an academic life one must have a love of writing. If you don't, forget it and do something else. Writing takes many forms: papers, reviews, talks, and grants. Grants are the lifeblood of an investigator, and good ones read like an elegant novel. Grant writing is a form of storytelling built around a series of technical observations that achieve biological significance after the ideas are rigorously tested for cohesiveness and validity. Many beginners fear grant writing, yet it is one of the best ways to stretch a hypothesis, shrewdly give it legs, and exercise creativity. Federal dollars are the backbone of our investigative enterprise today, even as the degree of competitiveness for funding variably cycles over succeeding decades. This periodicity is inevitable and challenges all of us to improve on the questions we ask; time has shown that the peer-review process for allocating grant dollars rarely fails to fund a really good idea. High-quality science thrives in a meritocracy—we just have to learn to be meritorious.

For most young medical students and residents, primary career choices fall to patient care, biomedical investigation, or industry, and for graduate students, the academic laboratory, industry, or teaching. Each of these choices creates its own special wrinkles; every busy scientist, clinician, or educator has issues, and much in this book deals with finding that special equilibrium between career and personal life. The women working at Vanderbilt address this balance spectacularly, and it would be a mistake to assume a life of scientific inquiry forces you to choose between family and career; rather, it is all about finding a clever way to manage both.

One critical attribute of the contributors to this anthology was their skill in time management. Most revealing was how many saw research life as more flexible than other career choices. For some, careers at the bedside totally immersed in patient care seemed more complicated than work at the bench, and returning to patient care no sinecure for insecurity when experiments reached an impasse. Exercising the intellect, traveling to interesting places, hard work in the laboratory, and having colleagues throughout the world are all perquisites of a gregarious research life. To participate ably, all of us need help from our university, day-care services, and personal support groups.

Nearly all the women in this book were married (one myth dispelled), most had children (another myth dispelled), but virtually all who married chose incredibly supportive spouses. For scientists who marry, one message is clear: find a spouse who recognizes the importance of your career and is willing to share responsibility for the work of living and raising children.

In Homeric mythology, Ulysses on his way to the Trojan War appointed Mentor to stay behind and educate his son, Telemachus, for life befitting a leader. Mentorship from that time forward has been a watchword for receiving a longitudinal education in career development. This special guidance usually requires several traits in a mentor . . . providing good counsel, prudent restraint, and practical insight.

Finding a mentor with whom you resonate is the first step. Mentors need to have a track record of working on meaty problems that interest. They need a sufficient amount of research experience to guide their mentee through a puzzle neither understand. Mentors also need professional gravitas and the time to be good problem solvers—and through these personal experiences, most mentors become lifelong friends.

Although much is written on the subject of selecting a mentor, mentorship is not always about matching gender; for practical reasons one has to be open to any good opportunity. Even Telemachus occasionally got into troubles that Mentor couldn't deal with and the goddess Athena had to step in to straighten things out. Some have argued that one of the barriers to a successful career for women in science is the lack of older women who can advise; this may be true at some level, but the stories here suggest a different message. One cannot help but be struck by the large number of men who successfully guided the careers of these female authors. Maybe expediency played a role or all were just lucky, but in the times in which we live, we work with what we have. Man or woman, if your mentor shows lasting interest in your success, then you have chosen wisely.

Finally, do we now know why women in large numbers are not joining in the work of discovery? Not really. This anthology only describes a special group who found their way, and their success forms the core of a small, and we hope, growing percentage of future scientists. Surprisingly, their route to making science a career was not always a direct one. In some cases, providence skillfully redeployed them toward academic life, and happily the system let them matriculate. Is access a critical barrier? I think anyone would say we could do better. The whole subject would benefit from a root-cause analysis. We need more women who are scientists on admissions committees and in national leadership roles; we need a closer look at women who are discouraged from training in science before they choose a discipline for advanced education. A love of science is not the only value in choosing matriculants to medical or graduate school, but it should not be a barrier, either. What these stories tell us is that if you want a career in research, there is a way to make it happen, and it can be yours.